Wayne O'Brien

Breakdowns in Controls in Automated Systems

Wayne O'Brien

Breakdowns in Controls in Automated Systems

VDM Verlag Dr. Müller

Impressum/Imprint (nur für Deutschland/ only for Germany)

Bibliografische Information der Deutschen Nationalbibliothek: Die Deutsche Nationalbibliothek verzeichnet diese Publikation in der Deutschen Nationalbibliografie; detaillierte bibliografische Daten sind im Internet über http://dnb.d-nb.de abrufbar.

Alle in diesem Buch genannten Marken und Produktnamen unterliegen warenzeichen-, marken- oder patentrechtlichem Schutz bzw. sind Warenzeichen oder eingetragene Warenzeichen der jeweiligen Inhaber. Die Wiedergabe von Marken, Produktnamen, Gebrauchsnamen, Handelsnamen, Warenbezeichnungen u.s.w. in diesem Werk berechtigt auch ohne besondere Kennzeichnung nicht zu der Annahme, dass solche Namen im Sinne der Warenzeichen- und Markenschutzgesetzgebung als frei zu betrachten wären und daher von jedermann benutzt werden dürften.

Coverbild: www.purestockx.com

Verlag: VDM Verlag Dr. Müller Aktiengesellschaft & Co. KG
Dudweiler Landstr. 125 a, 66123 Saarbrücken, Deutschland
Telefon +49 681 9100-698, Telefax +49 681 9100-988, Email: info@vdm-verlag.de
Zugl.: Fairfax, George Mason University, Diss., 2006

Herstellung in Deutschland:
Schaltungsdienst Lange o.H.G., Zehrensdorfer Str. 11, D-12277 Berlin
Books on Demand GmbH, Gutenbergring 53, D-22848 Norderstedt
Reha GmbH, Dudweiler Landstr. 99, D- 66123 Saarbrücken
ISBN: 978-3-639-08727-7

Imprint (only for USA, GB)

Bibliographic information published by the Deutsche Nationalbibliothek: The Deutsche Nationalbibliothek lists this publication in the Deutsche Nationalbibliografie; detailed bibliographic data are available in the Internet at http://dnb.d-nb.de.

Any brand names and product names mentioned in this book are subject to trademark, brand or patent protection and are trademarks or registered trademarks of their respective holders. The use of brand names, product names, common names, trade names, product descriptions etc. even without
a particular marking in this works is in no way to be construed to mean that such names may be regarded as unrestricted in respect of trademark and brand protection legislation and could thus be used by anyone.

Cover image: www.purestockx.com

Publisher:
VDM Verlag Dr. Müller Aktiengesellschaft & Co. KG
Dudweiler Landstr. 125 a, 66123 Saarbrücken, Germany
Phone +49 681 9100-698, Fax +49 681 9100-988, Email: info@vdm-verlag.de

Copyright © 2008 VDM Verlag Dr. Müller Aktiengesellschaft & Co. KG and licensors
All rights reserved. Saarbrücken 2008

Produced in USA and UK by:
Lightning Source Inc., 1246 Heil Quaker Blvd., La Vergne, TN 37086, USA
Lightning Source UK Ltd., Chapter House, Pitfield, Kiln Farm, Milton Keynes, MK11 3LW, GB
BookSurge, 7290 B. Investment Drive, North Charleston, SC 29418, USA
ISBN: 978-3-639-08727-7

Acknowledgements

I would like to thank my wife and three sons for their important encouragement over the years as I completed my doctoral program. My wife provided editorial suggestions and showed me patience and support throughout what became a long project, when we could have been doing things that were arguably more fun. The resulting dissertation comprises the bulk of this book. Dr. Rine gave me technical support that was essential to completing my dissertation as well as encouraging me to continue. Dr. Rine and Dr. Shortle were kind enough to coach me in how to briefly summarize and present the details contained in my dissertation. Dr. Sage was steadfast in remaining as my adviser for the duration, serving very effectively in that capacity. Finally, I received support from my employer during the entire program, in terms of tuition and recognition.

Table of Contents

List of Tables

List of Figures

Chapter 1: Background

Overview

Breakdowns in Controls in Automated Systems (*Breakdowns*) considers the general

problem of failing to satisfy organizational control requirements in automated systems.

By implication, *Breakdowns* is concerned only with those automated systems that have

an impact on organizational control requirements. What are organizational control

requirements (control requirements)? What is the significance of not meeting (satisfying)

them (Why is the topic important? Why is the topic interesting?)? What is the nature of

control requirements in automated systems? Is there anything distinctive about such

control requirements compared to those for manual systems (Why is the topic

challenging?)? This section is intended to answer such questions by providing a

background for control requirements and examples of what happens when control

requirements are not met, while highlighting the nature of control requirements in

automated systems.

What are control requirements? Nearly a hundred years ago, Henri Fayol described a

management process that continues to underlie current schools of management thought

[76]. Fayol's management process consists of five components: plan, organize,

command, coordinate, and control. This management process applies to military as well as civilian enterprises (the names of two of the components even appear in the familiar military term *Command and Control, or C2*). The following are general definitions of the control component of the management process:

> It is management's responsibility to exercise control .... The major reasons for exercising control are (1) to provide reasonable assurance the (organization's) goals ... are being achieved (2) to mitigate the risk that the enterprise will be exposed to some type of harm, danger, or loss (including loss caused by fraud or other intentional and unintentional acts) and (3) to provide reasonable assurance that certain legal obligations are being met [89].

> Within command and control, *control* is the regulation of forces and battlefield operating systems to accomplish the mission in accordance with the commander's intent. Control allows commanders to direct the execution of operations to conform to their commander's intent [240].

While the first definition relates to Accounting Information Systems (AISs) and the second relates to military Command and Control Systems (C2Ss), they capture the same essential concept – increasing the likelihood of reaching the goals of the organization (embodied in the commander in the military definition) by incorporating control.

The term *internal control* is often used in management and accounting literature for the control component, emphasizing control within the organization. The following are definitions of the control component using the term internal control:

> Internal control is a system of integrated elements – people, structure, processes and procedures – acting together to provide reasonable assurance that an organization achieves ... its ... goals [89].

> An internal control system that is working well can provide a foundation to help managers operate the firm effectively and efficiently, to issue reliable financial reports, and to meet the corporation's legal obligations [142].

Internal control affects all levels of the organization, so internal control is often

categorized to indicate the level at which control is exercised. For example, Gelinas [89]

provides the following hierarchy:

- The Control Environment

  Overall policies and procedures that demonstrate an organization's commitment to the importance of control

- Pervasive Control Plans

  Address multiple goals and apply to all applications

- Application Control Plans

  Relate to specific AIS subsystems or to the technology used to implement the subsystem

More generally, the control model for the management process relates to the first-order

feedback (closed-loop) system of cybernetics or systems theory [89].

Breakdowns views control requirements (or organizational control requirements or

internal control requirements) as those requirements imposed on an organization in order

to comply with – that is, to accomplish the purposes described in – the above definitions

of control and internal control. For brevity, such compliance is often stated simply as

providing *adequate internal control* (see examples below in this section) or, especially in the context of automated systems, an *adequate system of internal control.* Failing to satisfy control requirements results is an inadequate system of internal control. For readability, the adjectives *adequate* and *internal* are often dropped to refer simply to *control.* Further, when appropriate for the context, e.g., for a fine level of granularity, the plural of *control* is used to indicate that multiple specific *controls* are involved.

What is the significance (importance) of not meeting control requirements? In other words, what are the consequences of not providing adequate internal control? A series of high profile scandals and failures has led to a corresponding series of legislative and management responses related to controlling government and private enterprises. For example, the Foreign Corrupt Practices Act (FCPA) of 1977 was a response to illegal payments made by American companies in dealing with foreign governments during the Watergate era [112]. The 1992 report of the Committee of Sponsoring Organizations (COSO) of the National Commission on Fraudulent Financial Reporting (Treadway Commission) was a response to the savings and loan failures of the 1980's. More recently, the Enron and WorldCom failures led to legislation requiring stricter punishment of key corporate personnel responsible for companies with ineffective control environments – the United States' Public Company Accounting Reform and Protection Act of 2002 [224].

While the Enron and WorldCom failures resulted in losses of several billion dollars, they were dwarfed by the failure of federal agencies to account for their funds, such as the failure of the Department of Housing and Urban Development in its 1999 financial audit to account for fifty-nine billion dollars [183]. The 1990 Chief Financial Officers Act requires all twenty-four primary federal agencies to have audited financial statements. In fiscal year 2002 (ending September 30, 2002), auditors were unable to determine if the information was reliable in the financial statements for a third of the twenty-four federal agencies; auditors were unable to determine if the information was reliable in segments of the financial statements for four of the remaining sixteen agencies. Beyond such high-profile financial scandals and failures, recent news focused attention on command and control in New York City for first responders during and after the 9/11 terrorist attacks on the World Trade Center and on death caused by friendly fire (fratricide) during the ensuing wars in Afghanistan and Iraq.

The financial scandals, failures, and loss of life noted above represent breakdowns in the respective organizations' systems of internal control, as indicated by the emphasis on improving such control in the resulting legislation and investigations. E.g., the FCPA "requires companies whose securities are listed in the United States ... to devise and maintain an adequate system of internal accounting controls [224];" COSO identified a control model with five interrelated areas of internal control; the proceedings on one of the friendly fire incidents in Afghanistan highlighted potential failures in the C2S.

What is the nature of control requirements in automated systems? Is there anything distinctive (interesting, challenging) about such control requirements compared to manual controls? Coincident with breakdowns in control related to the above financial debacles and events related to loss of life, organizations have relied increasingly on automated systems for accounting and operations (e.g., electronic commerce). The accounting and auditing professions have responded with standards for developing such automated systems that have adequate control (e.g., the Control Objectives for Information and Related Technology (COBIT) developed under the auspices of the Information Systems Audit and Control Foundation, in response to COSO).

In military organizations, the growing recognition of the importance of control was reflected in the emergence during World War II of the phrase *Command and Control* (C2) in place of simply *Command* [240]. Military organizations also have come to rely increasingly on automated systems, including Command and Control Systems. "Since 1977 (coincident with the FCPA), C2 has been considered one of the battlefield operating systems [240]." The military parallel to electronic commerce is *Network-Centric Warfare (NCW)*.

One of the reasons the emphasis on control has paralleled the growing reliance on automated systems, whether in military or civilian organizations, is the radical change in the nature of controls in an automated system compared to manual systems:

- Computer processing is impossible to observe because it takes place within a box. By contrast, manual data processing occurs in plain sight and may be simply observed.
- Most computers and computer programs are too complex, even at their most fundamental level of operation, for anyone but

> specialists to understand. In contrast, almost anyone observing a
> manual data processing facility can understand what is happening.
> - Much computer output is in the form of magnetic storage,
>   impossible to inspect unless it is converted to paper or video
>   display. In contrast, manual data processing produces a readable
>   record.
>
> The effect of these changes on auditors was to (1) remove the well-
> understood manual data processing environment and (2) take away the
> audit trail – the documentation that describes a transaction's history [229].

The increasing speed at which transactions occur and the increasing distance over which

they occur are also characteristics of automated systems that contribute to this radical

change in the nature of controls. Further, the transformation from manual systems to

electronic commerce enabled by automated systems "… strips transactions of social

context. Control in e-commerce must be able to cover transactions without social context

among a broader set of agents who come in direct contact [230]." This radical change in

social context, in turn, introduces the consideration of organizational and cultural

knowledge (see below in this section). Considering just AISs and military C2Ss as

examples, manual control developed over centuries in the former and millennia in the

latter. The following examples show how increasing automation and technology change

the nature of control:

- If armies fought only with swords, fratricide would not be an issue. Adding

  spears, then arrows, rifles, and cannons made fratricide increasingly likely.

  Automated systems enable more advanced weapons technology. Speed, distance

  between shooter and target, and destructive power continued to increase. These factors made the nature of control in modern weapons systems change so completely that a separate systems control area emerged, Identification Friend-or-Foe (IFF) systems, providing a system of internal control to reduce fratricide.

- In AISs, a comparable example is in the rapid growth of electronic funds transfers. Transactions that use currency in a single physical location can be controlled by relatively infrequent counts of money compared to paper receipts. Now, with electronic commerce enabled by automated systems, 1.7 trillion dollars are transferred electronically between banks in the United States daily [37], making manual counting at infrequent intervals clearly an inadequate internal control.

In both examples, technological advances, including automation, rendered existing manual controls inadequate. From the perspective of controls, the evolution of technology created a problem, rather than a solution, upon the transformation from manual to automated processes, for the initial state of the controls in the related processes. Following the transformation, after observing the inadequacy of controls, stakeholders can use technology for the solution, e.g., IFF systems. *Breakdowns* will analyze the general problem and the constrained problem (see Constrained Problem section) from this perspective – where technology first has a negative impact on controls that then drives the application of technology to make the controls adequate once again. Consequently, *Breakdowns* does not distinguish whether the technology that may have a negative impact on controls is or was push- or pull-driven, nor does it consider such issues as Technology Readiness Levels [174], because the technology is already in place. As a side note, if *Breakdowns* were considering how to avoid negative impacts of new technology before implementing such technology, Technology Readiness Levels would be useful only if they included future ethnographic readiness as a criterion. By way of contrast, another perspective would be to assume technology could improve controls without having first noted an inadequacy in the control system caused by evolving technology.

To respond to this changing nature, either controls must be added to the existing systems or new systems developed. Adding or changing features in an existing "…system can easily be 3 times as expensive and take more than twice as long as writing …[136]" a new system. Moreover, because of the difficulty of updating aging software safely, the potential defects when updating old software can be 3 times higher than for new software…[136]" Such direct cost considerations added to the cost of inadequate internal control (e.g., the scandals and failures noted above) encourage the designing of adequate internal control into a system from the outset. There is also a recognition that the cause of failure begins at the beginning.

> When systems failures or disasters occur, blame is often placed on system operators, such as control room staff, pilots, drivers, or maintenance staff, for not acting in accordance with procedures or for not diagnosing a problem quickly enough. In investigations following a major failure, a common approach is to focus simply on the operational activities as the most likely cause.

>Instead, the author suggests that the investigation needs to delve further back into the project life. Original requirements, system design, development activity, and testing are all valid areas for review as is the control over system maintenance [13].

If control requirements are not included from the beginning with other requirements for the system, the risk of overlooking their interdependence with other requirements increases. Overlooking such interdependence can lead to an inadequate system of internal controls by omitting requirements or introducing adverse interactions that derive from that interdependence.

The desire to build control into automated business and military systems from the outset strongly suggests a consideration of the processes for developing such systems. A goal of systems development is to apply an "… overall set of technologies that simplify complexity and minimize a natural human tendency to make errors while performing complex tasks [136]." But how much effort in this regard is needed for adequate control?

>First is the complaint that *controls make the system more complex.* True, incorporating controls probably requires more analysis, more design, and more code, but doing so results in a system that does more things. I'm all for recommending that you plan to install the minimum set of controls necessary for them to be effective, but what's the value of a simple system *without* control? This complexity argument arises because when it becomes necessary to retrofit controls into existing systems, complexity can be a problem. But when controls are designed into a system from the start, the amount of complexity will be less, and it will be more than made up for by increased accuracy and reliability [229].

The starting point for processes to develop systems is determining the requirements for the system to be developed – what does the system need to do? Requirements involve both *explicit knowledge* and *tacit knowledge.* Explicit knowledge (see Explicit

Knowledge, Appendix B) may be active (embedded in human consciousness [170 and

266]) or passive (*information* [170]). Tacit knowledge (see Tacit Knowledge, Appendix

B) is implicit knowledge that is experience-related, applied unconsciously, or taken for

granted – see example of the cook below; [188 and 195]). Such explicit knowledge must

be captured – that is, identified, extracted, and stored as artifacts (see Figure 7 for

examples) for use in developing the system (see Background of the General Problem

section).

In addition to explicit knowledge about requirements, tacit knowledge of requirements

may be involved in two ways. First, people may perform activities as routines that they

give little or no thought to, especially the reasons for performing them. The following

story illustrates this. Whenever a certain cook prepared a roast, he cut it in half and

placed it in two pans. After observing this a number of times, one of his assistants asked

him why. He said his grandmother had always done it that way. When the cook's

grandmother paid him a visit, the assistant took the opportunity to ask her why she

cooked roasts that way. She said that the pans she had were too small to hold the entire

roast. If a new automated system were based on the cook's tacit knowledge (in his case,

performing a routine based on the implicit assumption that there was good reason to do

so in a certain way), an unnecessary step would be included – always cutting the roast in

half.

Second, an automated system to be replaced by a new system may perform functions that are not known or incompletely known to the people using the system, perhaps because the automated system to be replaced is undocumented. If such unknown or incompletely known functions are not performed by the new system, their absence may go undetected until the effect becomes visible in the results produced by the new system, e.g., an erroneous calculation or a certain type of error that had been detected by a missing function. I have worked with government systems – used in the management of multi-billion dollar programs – that had thirty-year old embedded algorithms that were undocumented and whose use maintenance programmers could not determine. To err on the side of not having undetected problems, the algorithms were routinely carried over to new systems.

As the two examples show, the consequence of tacit knowledge may be either to include something that is not needed or to omit something that is. Likewise, failure to capture (see below in this section and the Background of the General Problem section) all relevant explicit knowledge in the requirements for an automated system may result in inclusion of unnecessary features or omission of necessary ones. Stated in another way, some "requirements" may be either missing or unnecessary, whether because of tacit knowledge or failure to capture relevant explicit knowledge. Perhaps the most common causes of not capturing all relevant explicit knowledge are not identifying all relevant sources (e.g., not talking to the grandmother) and not extracting such knowledge (e.g., not asking the grandmother the right question). These factors contribute to causing requirements defects, one of the larger categories of software error [136].

Breakdowns applies knowledge management theory to the issues of tacit versus explicit knowledge and to the organizational and cultural knowledge that is related to the transformation from control based on manual transactions to control based on electronic transactions. That is, knowledge management theory is useful in converting tacit knowledge to explicit knowledge. Tacit knowledge may be embedded in organizational and cultural routines, which may themselves be embedded in automated systems.

There are certain common issues of automated systems that developers must deal with, regardless of the problem domain or application area. *Breakdowns* focuses on the common issues of failing to provide adequate internal control in the automated systems for the problem domains of Accounting Information Systems and Command and Control Systems, with inferences that can be applied to a broad range of problem domains or application areas.

As noted above, the need to build adequate internal control into a system from the outset encourages consideration of the processes used for developing systems in general and the starting point for such processes is determining the requirements for the system to be developed. Consequently, analysis of such processes and development and application of a new process are major parts of *Breakdowns*. This new process encourages two key outcomes:

- o The capture of relevant requirements – avoiding both the inclusion of unnecessary features and the omission of necessary ones – for adequate internal control to be built into a new system

- o A means of adding or changing controls (after the new system is in operation) that mitigates the problems noted above when adding or changing features in existing systems

Organization of *Breakdowns*

The remainder of *Breakdowns* continues with the sections Problem Discussion, Solution Approach for the Constrained Problem, Validation, Research Approach, Survey of Related Literature on prior research, and the Prototype used to demonstrate the solution.

Problem Discussion

General Problem

Using the key points discussed in the Overview section above, the general problem stated there may be qualified as follows: failing to satisfy control requirements in systems where longstanding social and cultural issues involving humans and organizations (response to radical change in the nature of controls) are confronted with evolving technological changes (increasing use of automated systems). For brevity, the longstanding social and cultural issues involving humans and organizations will be referred to as the *ethnographic context* (or *conditions*); the evolving technological changes as the *technological context* (or *factors*).

In order to examine the confrontation between the ethnographic and technological
contexts in view of the issues raised in the literature (see Survey and Review of the
Literature of Prior Research section), *Breakdowns* decomposes the two contexts into
three interrelated technological factors interacting with three ethnographic conditions (see
Background for the Constrained Problem section and Table 1, Figure 3, and Figure 4).
The three technological factors considered are: proportion of automated versus manual
systems, proportion of automated systems that are integrated with each other, and
electronic or network-centric environments (see Figure 3 and Figure 4). The three
ethnographic conditions are longstanding reliance on physical records and barriers,
periods within which humans can react, and interpersonal relations (see Table 1). The
section, Background for the Constrained Problem, analyzes how the negative impact of
technology, noted above in the Overview section, results from the interaction of these
conditions and factors. The systems to which this ethnographic-technological context
applies will be referred to simply as systems throughout *Breakdowns*. Because
automated systems typically are software intensive, references to systems will assume
that the systems are software intensive. More specifically, beyond the general discussion
of control for background and context, the controls of particular interest in *Breakdowns*
for such systems are those implemented through prescriptive statements (see Internal
Controls section and Table 20 and Table 21) embedded in software code or through
prescriptive rules maintained externally to the software code (see Bifurcated Architecture
to Encompass Commonality and Variability section).

Control is one of five major components – Plan, Organize, Coordinate, Command, and

Control – of the standard management process found in the major schools of management

thought (see Figure 1 [76]). *Breakdowns* uses the term *control* to refer to the control

component of this management process and deals only with that component. As a

component of the overall management process for an enterprise, control encompasses all

levels of control, e.g., strategic, operational, and tactical.

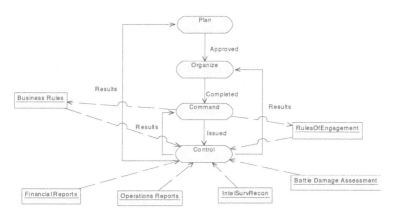

Figure 1: Management Process

Breakdowns examines the causes of failures in satisfying control requirements – resulting in an inadequate internal control – under the circumstances described above in the context of experience and the theoretical context of knowledge management (see Appendix B, Narratives and Time, and the Background of the General Problem section). As implied above, knowledge management theory examines the distinction between explicit knowledge and tacit knowledge embedded in human routines, organizational culture, and existing systems (see Appendix B, Explicit Knowledge and Tacit Knowledge). In knowledge management terms, control requirements exist as both explicit and tacit knowledge [66, 170, 188, 193, 194, and 195]. *Stakeholder* (see Stakeholder, Appendix B) control requirements must be extracted from both explicit and tacit knowledge to reduce the number of missing requirements (incompleteness) as well as to reduce the number of unnecessary requirements. One of the new concepts introduced in *Breakdowns*, *domain rules* (see Internal Controls section and Domain Rules section), offers a means to reduce incompleteness in requirements by bounding the problem domain.

Tacit knowledge of control requirements is a challenge for requirements elicitation, but *Breakdowns* is not generally concerned with requirements elicitation. In general, *Breakdowns* is concerned with how best to capture requirements after they have been elicited, including those embedded in tacit knowledge (such as domain rules), as development artifacts (natural language narratives, graphical representations, or software code in various forms describing the desired and/or current systems – see Figure 6 for examples) in such a way that they are explicitly known and readily accessible to authorized stakeholders throughout the lifecycle of the system.

Notwithstanding this general concern of *Breakdowns*, because of the central role of domain rules in the proposed solution and their often-tacit nature (see Procedure section, step 1.2), *Breakdowns* identifies where knowledge management theory is valuable in identifying domain rules (see discussion of Patriotta's knowledge management framework in the next section). The remainder of *Breakdowns* will generally use the term *control needs* (or stakeholder control needs) rather than control requirements (see p. 32), reflecting the need to capture tacit as well as explicit requirements, because requirements in the literature (see Survey and Review of the Literature of Prior Research section) are often assumed to be those explicitly provided by stakeholders and in documents available to developers, such as an external solicitation or internal request. That is, there are control needs that must be met by the system of control in order for it to be adequate, even if such control needs are not identified as requirements (see discussion in the next section regarding the articulation of control needs as artifacts). For brevity, the phrase *meeting control needs* will be used to mean providing an adequate system of internal control (see p. 3).

Background of the General Problem
Organizational controls (with humans in the loop) historically relied heavily on ethnographic conditions of interpersonal relationships, physical records and barriers, and

time intervals for feedback sufficient for humans to react. Controls designed to be valid, based on explicit knowledge of control needs in one time period, become routine if they are successful over time. The routinization of controls implies their success. Lack of success would have prevented such routinization 188]. *Breakdowns* uses Patriotta's knowledge management framework to look at the knowledge management context of controls through the lenses of *breakdowns*, *narratives,* and *time* (see Breakdowns, Narratives, and Time, Appendix B; [188]), to shed light on control needs and obstacles to meeting them. Given the long histories of the ethnographic conditions on which controls have been based, the lens of time is especially useful (see Domain Rules Analysis section in Chapter 2 and the Procedure section in Chapter 3). Because both the common-usage meaning and the knowledge management meaning of breakdowns capture the meaning of a failure in terms of the general problem, the term *breakdown*, rather than *failure* generally will be used in the remainder of *Breakdowns*.

Routines (and the needs they meet) become a tacit part of the control *background* (see Background, Appendix B) in the current and following time periods, applied unconsciously to make controls effective [188]. Thus, the control background consists of layers of tacit knowledge from multiple time periods (see Appendix B, Background). The ethnographic conditions affect the control background, e.g., when a routine task tacitly assumes a sufficient time interval is available to reconcile accounts before fraud occurs. They also affect the explicit knowledge of those performing the routine, as when a commander presents rules of engagement at a briefing, assuming the availability of interpersonal relationships.

When evolving technology leads to systems that intentionally embed explicit knowledge for the purpose of carrying out tasks otherwise performed by people (replacing those tasks and related manual controls), the systems accelerate the conversion of explicit knowledge to tacit knowledge, adding to the conversion caused by routinization of human tasks. In practice, the system, viewed through the lens of time, becomes part of the background against which human actors carry out their routines in the *foreground* [188]. The system, as part of the background, has tacit knowledge embedded in it.

A similar phenomenon could occur for purely ethnographic reasons, e.g., someone might replace a manual control with another manual control, but *Breakdowns* is concerned with replacement by systems that do not capture the tacit requirements. There is a difference when one manual control is replaced by another, rather than by a system. People apply the new manual control against the control background, whereas systems do not, except to the extent that the tacit knowledge of control needs (requirements) embedded in the background has been made explicit and articulated as narratives and the narratives captured as development artifacts (the system development equivalent of articulating narratives as text in knowledge management – see Appendix B, Narratives) for development and maintenance of the systems. For brevity, these multiple steps for the transformation of stakeholders' knowledge – whether explicit or tacit – of control needs into development artifacts (including software code) will be referred to collectively as articulating control needs (or knowledge or requirements) as artifacts, or an equivalent phrase, e.g., requirements articulated as artifacts.

That is, systems replace not only the explicitly known and documented control needs, they also replace, *de facto*, the tacit undocumented or unused background knowledge, whether or not such background knowledge of control needs is articulated as artifacts.

Failure to articulate tacit knowledge of control needs (including that embedded in existing systems to be replaced by the new system) as artifacts, just as a failure to articulate explicit knowledge of control needs as artifacts, means no adequate control (either missing or weak as a result of incomplete understanding) will be implemented, causing a gap in control in the system. Thus, it is essential to extract tacit knowledge of control needs from the control background, not just elicit explicit knowledge from stakeholders.

Such failures to articulate knowledge of control needs as artifacts can occur during the initial development of systems, when control needs change, and when breakdowns are corrected (see Solution Approach for the Constrained Problem section). Changing control needs could arise for multiple reasons, including articulation of previously tacit knowledge as artifacts (see Appendix B, Narratives) and new responsibilities assigned to the system by stakeholders (see Appendix B, Adaptability and Extensibility). The resulting lack of artifacts causes breakdowns whenever an event occurs that is not properly handled because of the related weak or missing control (i.e., a control that was not implemented, or not properly implemented, because the control need was unknown or inadequately known to the developers).

When stakeholders observe symptoms of such breakdowns, they are forced to focus on recovering any embedded knowledge concerning the breakdown. Without information (see Information, Appendix B) articulated as artifacts that the stakeholders can convert to knowledge to correct the breakdown (reversing the entropic dynamics – loss of explicitness and organization over time and among developer disciplines such as analysts and designers – through which the knowledge became tacit –see Narratives in Appendix B), stakeholders are left with the lens of time (see Solution Approach for the Constrained Problem section and Appendix B, Time) to attempt recovery.

If stakeholders articulate the recovered knowledge as artifacts, including *lessons learned*, the information will be available for future breakdowns. Lessons learned consist of knowledge gained both positively (e.g., during development of a system) or negatively, from a breakdown. If they recover sufficient knowledge to identify the root cause, they can correct the root cause. Further, if they include the root cause as part of the lessons learned and apply the lessons learned they can avoid future breakdowns of a similar kind.

Otherwise, the new background, provided by the new system, modified by whatever corrective action was taken, remains unarticulated as artifacts in the following time periods, forgotten with the passage of time (receding farther into history; also see

Appendix B, Narratives and Time, [188]). Future breakdowns require greater effort and time (resulting in greater costs) to recover the tacit knowledge, about which stakeholders' memories are likely to fade, from the control background, which now has the additional layer of tacit knowledge embedded in the correction (see discussion of layers in the Background for the Constrained Problem section and Appendix B, Background).

Both initial development and later correction of breakdowns may therefore fail to produce sufficient artifacts for knowledge recovery, increasing future costs. The act of correcting breakdowns, whether or not recovered knowledge is articulated as artifacts, often leads to additional breakdowns [136]. Needs for new and modified controls could arise for reasons described above in this section. Evolving technology, for the reasons described in the Overview and this section above, can drive the need for new or modified controls. Further, evolving technology suggests a recurring need for new or modified controls. This suggests looking at the general problem over time from the following four views:

1. Initial control needs for new systems

2. Needs for new and modified controls

3. Uninterrupted integrity of existing controls

4. More timely and less costly responses to recurring changes in needs underlying new and modified controls

Formal Statement of General Problem

The preceding discussion leads to the following formal statement of the general problem:

Breakdowns occur in meeting control needs in systems, under the following conditions:

1. Evolving technology has led to systems that are:

    1.1. Software intensive

    1.2. Highly integrated

    1.3. Network centric

2. The systems were intended to replace human routines, established over long periods of time, that rely on:

    2.1. Physical records and barriers

    2.2. Time intervals for feedback sufficient for humans to react

    2.3. Interpersonal relationships

3. Explicit and tacit knowledge of control needs are not articulated as artifacts for the development, operation, and maintenance of the systems

4. The control background recedes farther to history

Importance of the General Problem

The evidence for the two problem domains included in the constrained problem (discussed below) suggests that the negative impact of evolving technology is sufficient to render controls in systems in both domains inadequate [10, 30, 50, 51, 59, 74, 83, 85, 87, 102, 135, 151, 177, 180, 196, 199, 205, 229, 231, 233, 244, 245, 246]. The wide divergence of the two domains suggests that the negative impact is likely in other problem domains.

Figure 1 shows that control consists of disseminating information from Command to Control (the dashed arrows from Command to the Business Rules and Rules of Engagement rectangles show examples of such information), from information sources back to Control (e.g., Battle Damage Assessment and Operations Reports), and from Control back to Command, Organize, and Plan for Coordination. The content of the information, especially of Business Rules and Rules of Engagement, is essential to the management process, because the content is not simply declarative, but rather imperative. In systems, this imperative content includes the means to enforce the rules. Control provides the means for ensuring that Command within an enterprise is executed and executed correctly (see the Internal Controls section and Appendix B, Controls). Indirectly, information flows from Plan and Organize, through Command, to Control. That is, an adequate system of control begins at the beginning, with planning and organization.

If the management process were to steer a ship, the control component would be the rudder; without adequate, valid control, the management process in any organization would be rudderless. Plans, organization, coordination, and commands and to steer the ship would be ineffective. For the civilian organization, the consequences include financial loss, bankruptcy, and loss of life (e.g., unsafe products, unreliable medical systems); for the military organization, national defense would be at risk, with obvious life-threatening or political implications.

Constrained Problem

The general problem is constrained by focusing the discussion in *Breakdowns* on two problem domains: accounting information and command and control (subsequent references to two domains will assume the accounting information and command and control domains). Two such widely divergent domains were chosen to emphasize the significance and generality of both the problem and the solution. The prototype (see Validation and Prototype sections) further constrains the general problem by representing one of the two domains and by dealing with a subcategory of controls (see Internal Controls section).

Background for the Constrained Problem

Both systems in the domains selected for the constrained problem, Accounting Information Systems (AISs) and Command and Control Systems (C2Ss) are particularly sensitive to the confrontation between longstanding ethnographic conditions and evolving technology described above. Controls in both domains developed over centuries, relying on personal interaction and manual procedures, creating many layers of tacit knowledge within the control background as explicit knowledge receded to history (see Appendix B, Meta-Artifact, Tacit Knowledge, and Time). Checks and balances in the management processes of both domains presupposed the control characteristics of the ethnographic conditions, relying heavily on the deeply layered control background.

The remainder of this section elaborates on the nature of this sensitivity in terms of the three ethnographic conditions and three technological factors listed above in the Formal Statement of General Problem section. In both domains, decades of automation have steadily attempted to complement, substitute for, or obviate manual controls. For example, Identification Friend-or-Foe systems (see the Importance of the Constrained Problem and Appendix B, Command, Control, Communications, and Computer Systems) in C2Ss, or in AISs, batch controls, check digits, data value constraints, and elimination of cash handling with electronic banking.

The overall risk born by an enterprise's systems increases as the reliance on them expands. Risk is defined as the product of the likelihood of a control breakdown and the impact of the breakdown. The breakdown may be caused by a missing control or weakness in the control (see the Background of the General Problem section).

Figure 3 and Figure 4 show graphically the impact on risk of the three technological factors as evolving technology overcomes the ethnographic conditions (see Table 1) and systems move into the northwest quadrants of those figures. As the proportion of an enterprise's activities that is automated increases, manual authorization and decision processes are replaced (see Table 1) and the need for automated controls increases, causing the risk attributable to such needed automated controls to increase. As the integration of these automated activities increases, the use of manual procedures and controls decreases even more (e.g., transactions pass directly from one application to another without intervening manual reconciliations or authorizations, for instance from Purchasing in Figure 2 to Cash Disbursement or from Space Reconnaissance to Understanding and Battlemanagement in Figure 5), causing the risk attributable to needed automated controls to increase further. A system in the low-risk quadrant of

Figure 3 would have fewer than half the ovals (use cases) in Figure 5 or fewer than half the boxes (applications) under the AIS in Figure 2 automated; of those, fewer than half would be integrated.

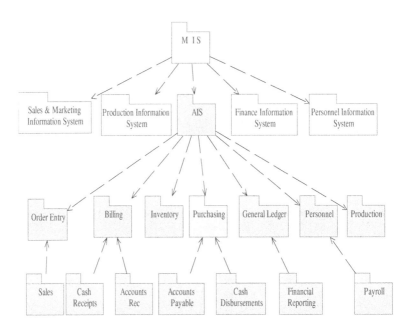

Figure 2: AIS Context

Using the halfway point as the low-risk threshold is only notional – there is no empirical

significance to using the halfway point as the boundary between low and medium risk.

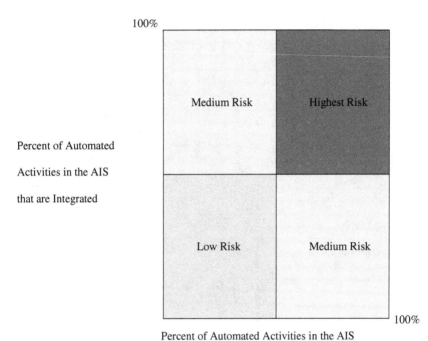

100%

Medium Risk | Highest Risk

Percent of Automated

Activities in the AIS

that are Integrated

Low Risk | Medium Risk

100%

Percent of Automated Activities in the AIS

Figure 3: Integration of Automated Activities in the AIS or C2S

100%

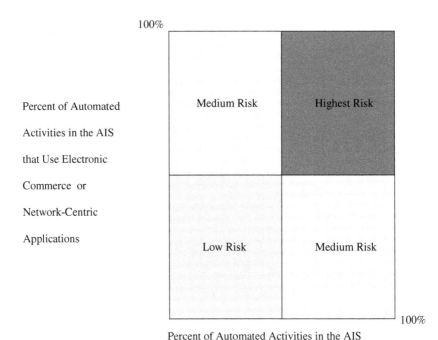

Percent of Automated

Activities in the AIS

that Use Electronic

Commerce or

Network-Centric

Applications

	Medium Risk	Highest Risk
	Low Risk	Medium Risk

100%

Percent of Automated Activities in the AIS

Figure 4: Use of E-Commerce by Automated Activities in the AIS or C2S

Similarly, as automation increases and the proportion of automated activities that depend

on electronic commerce or network-centric applications increases, additional manual

authorization and decision processes are replaced– e.g., check signing or personal interaction – increasing the reliance on automated controls and the risk still more. The risks stemming from increased reliance on electronic forms of authorization and identification are compounded by the speed and physical separation with which electronically transmitted transactions occur (see Table 1). A system in the low-risk quadrant of Figure 4 would have fewer than half the ovals (use cases) in Figure 5 or fewer than half the boxes (applications) under the AIS in Figure 2 automated; of those, fewer than half would rely on network-centric operations or electronic commerce.

Each of the above three migrations, corresponding to the three technological factors, out of manual controls that relied on the three ethnographic conditions, and into automated surrogates, is an example of embedding knowledge in systems. Unless the knowledge (both tacit and explicit) of control needs is articulated as artifacts, the knowledge recedes to history (or farther to history, in the case of tacit knowledge; also see Appendix B, Meta-Artifact, Tacit Knowledge, and Time) with the passage of time, making it extremely difficult to recover for any purpose, e.g., correcting breakdowns, training, or operations.

The risks described in conjunction with Table 1, Figure 3, Figure 4 arise from the deviation from the pre-existing ethnographic conditions and the control background (see Background of the General Problem section) on which control needs are based. Need is a key term in this context, because the need exists, but is not necessarily formulated as a requirement or met by the automated controls that replace the former controls, whether they were manual, semi-automated, or automated (see Background of the General

Problem section).

Table 1: Impact on Controlled Transactions

Impact on Controlled Transactions of Automation, Integration, and Network Centricity	
Reduced use of physical records and barriers	Less visibility because of structural changes, e.g., fewer paper documents, fewer reconciliations relying on human routines (e.g., adding machine tapes and signatures), locked doors, and physical inventories [89, 205, and 254]. There is substantial assurance when examining a physical record (e.g., a paper document) that it is authentic. This assurance may be far less in the case of electronic records, which may be rewritten without a trace, along with corroborating records to provide consistency with the basic alteration.
Reduced reaction time	More abstract, faster-occurring errors or irregularities because of less visible, totally electronic, realtime processes – audits traditionally were timed and planned to detect traditional errors or irregularities, which occur over time measured on a human scale like days or months
Reduced interpersonal contact	Less reliance on trust based on human interaction, such as personal relationships, face-to-face contact, and human audits to detect fraud, versus legal, technological, and analytical substitutes, such as trading partner agreements, pattern matching with neural nets, and statistical sampling 74 and 254].

Just as electronic commerce is the most recent factor in evolving technology to increase

the risks of control breakdowns in AISs, the concept of Network Centric Warfare (NCW)

emerging over the past several years, has increased such risks in C2Ss.

> Warfare takes on the characteristics of its age. NCW continues this trend
> – it is the military response to both the challenges and the opportunities
> created by the Information Age. Network Centric Warfare is to warfare
> what e-business is to business. The term Network Centric Operations
> provides a useful shorthand for describing a broad class of approaches to
> military operations that are enabled by the networking of the force. When
> these military operations take place in the context of warfare, the term
> Network Centric Warfare is applicable [177, p. iii].

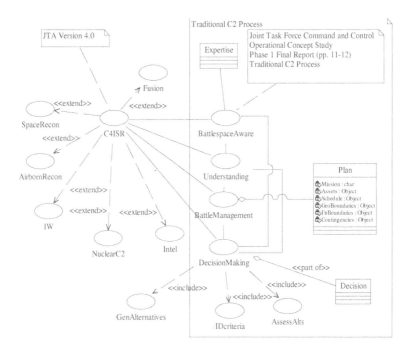

Figure 5: Command and Control Context

Figure 5 shows the context for Command and Control Systems. The characteristics and

subdomains of the C2 domain in Table 2 [135, p. 12] are becoming increasingly

automated, integrated, and network-centric, as are the subdomains of the C4ISR

(Command, Control, Communications , Computers, Intelligence, Surveillance and

Reconnaissance) domain in Table 3. Both domains, as suggested by Figure 5, are

becoming more integrated with each other:

> One concept being pursued to attain information superiority is known as
> NCW. The concept of NCW links sensors, communications systems, and
> weapons systems in an interconnected grid that allows for a seamless
> information flow to warfighters, policy makers, and support personnel
> [177, p. 1-1]

Command and Control (C2) may seem to be an abbreviated form of Command, Control,

Communications, Computers, Intelligence, Surveillance and Reconnaissance (C4ISR),

but C4ISR supports C2 (see Appendix B, Command, Control, Communications, and

Computers). In addition, other domains such as Weapon Systems (see Table 3) are being

integrated with the C4ISR and C2 domains, and increased integration is the goal [64, p.

1].

Table 2: Command and Control Domains and Subdomains

Command and Control Domains & Subdomains
Battlespace Awareness (subdomain)
Understanding
Decision Making (subdomain)
Decisions
Battle Management (subdomain)
Information Systems

Whenever manual procedures, including face-to-face contact and physical documents, are replaced in the C2 process, the factors affecting controlled transactions listed in Table 1 would apply. This is in fact the case in areas such as transmitting a commander's intent and Rules of Engagement (ROE), as they are transmitted instantaneously to all echelons, rather than through personal briefings and signed papers.

> The cognitive domain is in the minds of the participants. This is the place where perceptions, awareness, understanding, beliefs, and values reside and where, as a result of sense making, decisions are made. This is the domain where many battles and wars are actually won and lost. This is the domain of intangibles: leadership, morale, unit cohesion, level of training and experience, situational awareness, and public opinion. This is the domain where an understanding of a commander's intent, doctrine, tactics, techniques, and procedures reside. Much has been written about this domain, and key attributes of this domain have remained relatively constant since Sun Tzu wrote *The Art of War.* [177, p. 3-9]

> Developing the technologies and procedures that provide high confidence that information available to the warfighter is protected, available when needed, and can be trusted (includes capabilities for defensive information warfare and access control based on agreed security and need-to-know requirements). Focus is on (1) the ability to protect DoD information, systems, and networks from attack, (2) the capability to detect information warfare attacks in real-time, and (3) the ability to react quickly to ensure mission critical information is available, correct, and secure. [177, p. 10-14]

> The principle of security is also fundamental to military success. In today's military this translates into Information Assurance providing an uninterrupted flow of authentic communications and information. If the information processing or communications channels are compromised, or feared to be compromised, military success is imperiled [177, p. 3-19].

Table 3: Integration of C4ISR and Other Domains and Subdomains

Domain	Subdomain
C4ISR	Cryptologic
	Nuclear C2
	Space Reconnaissance
	Air Reconnaissance
Modeling and Simulation	
Combat Support	Automatic Test Systems
	Defense Transportation systems
	Medical
Weapon Systems	Aviation
	Ground Vehicles
	Missile Systems
	Missile Defense
	Munition Systems
	Soldier Systems

Formal Statement of Constrained Problem

The discussion in the preceding section leads to the following formal statement of the constrained problem:

Breakdowns occur in meeting control needs in systems under the following conditions:

1. The systems are in the domains of either the accounting information or command and control systems

2. Evolving technology has led to systems that are:

   2.1. Software intensive

   2.2. Highly integrated

   2.3. Network centric

3. The systems were intended to replace human routines, established over many centuries, that rely on:

   3.1. Physical records and barriers

   3.2. Time intervals for feedback sufficient for humans to react

   3.3. Interpersonal relationships

4. Artifacts (including lessons learned) supporting explicit knowledge of the control background are missing or unused

5. The control background recedes farther to history

Importance of the Constrained Problem

As noted in the Background for the Constrained Problem section above, the AIS is itself a general control, serving both legally and traditionally as the official system of record for the civilian enterprise, making it a focal point for controlling the management process for such enterprises. For the military enterprise, "command and control (C2) is an essential element of the art and science of warfare. No single specialized function, either by itself or combined with others, has a purpose without it [240]." Because of the central role of

AISs and C2Ss in their respective enterprise's system of control, the constrained problem amplifies the importance of the general problem.

Solution Approach for the Constrained Problem

The General Problem section above identified four views of the general problem:

1. Initial control needs for new systems
2. Needs for new and modified controls
3. Uninterrupted integrity of existing controls
4. More timely and less costly responses to changes in needs underlying new and modified controls

The solution approach considered throughout this section reflects these four views. The means of avoiding breakdowns in meeting control needs in the two domains (subsequent references to control needs will assume the two domains) presented in this section and described in later chapters are technological, but there is a potential non-technological effect that might contribute to the solution. There could be a cybernetic effect as people who interact with the system become increasingly familiar with the systems that are subject to the three technological factors listed above. However, the extent to which this cybernetic effect would mitigate the general problem is beyond the scope of *Breakdowns*.

Solving the constrained problem in terms of the theoretical knowledge management context, as well as the pragmatic ethnographic-technological context, described in the

Problem Discussion, requires understanding both the explicit knowledge of stakeholders and the implicit (tacit) knowledge of the control background.

Initially, the concern is to build adequate control into a new system. During the lifecycle of the system, the concern shifts to responding quickly and economically to new and changed control needs, while maintaining the integrity of the existing controls. When such control needs are not responded to quickly enough, breakdowns occur, disrupting the routine functioning of the system.

Based on historical evidence [7, 41, 47, 48, 49, 58, 62, 88, 155, 198, 200, 200, 211, and 234], lifecycle-centric system development in practice suffers from disconnects (gaps) among the phases[1] of software development and developer disciplines throughout the lifecycle, leading to increased costs and defects (see Contrasting the Meta-Artifact with

---

[1] The number and names of development phases differ among processes, but for brevity, this dissertation refers to the phases *requirements, analysis, design,* and *implementation,* which are also the names of the first four core workflows (development activities) within an iteration in the Unified Software Development Process (USDP [22]; see Iterative, Incremental Methods section, where phases are groups of iterations), which is the underlying process used by the procedure (see Procedure section and Table 15) for applying MAP

Artifacts of Other Processes section for a detailed discussion) for the system, including its controls. The impact of these gaps (see Contrasting the Meta-Artifact with Artifacts of Other Processes section for a detailed discussion of gaps and Figure 12 for a graphical representation) is compounded for systems that must interact with other systems in their respective domains (AIS and C2S for the constrained problem). That is, the gaps among phases and disciplines in developing a single system are increased by the completely different development cycles of separate systems (see Meta-Artifact Process section for a detailed discussion of these gaps).

In knowledge management terms, the gaps result from missing narratives and/or the practical means of converting narratives into explicit knowledge, because of the form (artifacts) the narratives are in or the tools available. Breakdowns occur when developers for the next phase try to apply the knowledge and routines of their disciplines against a tacit background that differs from that of the preceding discipline (semantic gaps, see Contrasting the Meta-Artifact with Artifacts of Other Processes and Figure 12). That is, the explicit knowledge as well as the tacit knowledge of the disciplines differs. Over time, the explicit knowledge from the previous phase (or another discipline) recedes to history (see Appendix B, Meta-Artifact, Tacit Knowledge, and Time) in the background of that phase, making it less accessible to the next phase with the passage of time (temporal gaps).

The lenses of time, breakdowns, and narratives (see Appendix B) can be applied to reverse the entropic dynamics of the semantic and temporal gaps. The breakdown focuses stakeholder attention on the need to recover the embedded knowledge. Developers can reconstruct the tacit knowledge through narratives that retrace the temporal sequence of actions (reverse the knowledge creating dynamics – see Appendix B, Narratives), some of which may have been articulated in text or other system artifacts. This is a costly, duplicative effort encountered in system maintenance activities performed on operational systems that lack current, comprehensive documentation. Due at least in part to such circumstances, system lifecycle maintenance accounts for the majority of lifecycle costs [136]. Changes to software source code for maintenance are costly and have a high probability of introducing defects into the system [136] (see Bifurcated Architecture to Encompass Commonality and Variability section).

The preceding discussion in this section suggests that avoiding breakdowns in meeting control needs under the pragmatic ethnographic-technological context and the theoretical knowledge management context of the general problem requires a systems development process that has the following capabilities:

1. Comprehensively articulates stakeholders' explicit and tacit knowledge as narratives for conversion into development artifacts that are retained as information that can be converted back to active knowledge to assure adequate

understanding and proper implementation of control needs during the complete lifecycle of the system

2. Reduces or eliminates the gaps among development disciplines and development phases during development , and among systems in a domain during operations and maintenance

3. Considers other existing systems and future systems in conjunction with the system under development because of both the explicit and tacit knowledge contained in the other systems, on which the system under development may depend, especially when they must interact with each other. This in turn requires a comprehensive consideration of the domain in which the systems reside.

4. Reduces the need to change software source code for maintenance resulting from breakdowns or changing control needs, with resulting reductions in costs and defects

Domain analysis [7] is an established approach for analyzing requirements for multiple systems in a domain. Object-Oriented Domain Analysis (OODA) extends domain analysis by using Object-Oriented technology. Thus, OODA offers a starting point for providing capability 3 (references to capabilities in the rest of this section are to the immediately preceding list of four capabilities). Extensive research supports Object-Oriented Domain Analysis (OODA) as an approach to examining an entire domain (see Domain Rules Analysis section and [7, 49, 94, 200, 200, and 202]). Research also identifies limitations for OODA as currently defined and practiced. As discussed in detail in Chapter 2, one of the difficulties in the application of OODA lies in defining the boundaries of the domain [7 and 202]. Another is how to stay focused on the problem domain sufficiently to avoid moving prematurely to the solution space, before adequately considering all of the requirements of the problem space, including control needs. This suggests the need for augmenting OODA in order to efficiently and effectively define a domain so that it may be comprehensively analyzed (capabilities 1 and 3).

Once the domain has been analyzed to provide sufficient completeness to adequately meet all stakeholders' control needs, capability 4 requires the construction of prescriptive instructions (see Internal Controls section and Table 20 and Table 21) to verifiably and validly implement the control needs and maintain the integrity (including architectural integrity – see Architecture Centricity section in Chapter 2) of the prescriptive instructions over the complete lifecycle of the system if breakdowns occur or control needs change. The artifacts stored as information for capability 1 above must be related so as to provide a valid and accessible understanding of the stakeholders' control needs whenever they are converted to active knowledge during the complete lifecycle of the system, including all of the phases of development, regardless of which discipline is creating or using the artifacts (capability 2).

This again involves the lifecycle-centric systems development process, with emphasis on three persistent problems in software development technology that affect iterative and incremental system lifecycle integrity, development time, and cost: inadequate software reuse, reliability, and interoperability.

Therefore, an informal research hypothesis is that Object-Oriented Domain Analysis (OODA), augmented with the four capabilities listed above, would provide an improved process for developing systems that would be a potential solution to the constrained problem. The improved process is referred to as the Meta-Artifact Process (MAP). With this in mind, Chapter 2 discusses three interrelated enabling extensions (for brevity, referred to simply as extensions) to augment OODA. One is Domain Rules Analysis, which primarily addresses the problem of completeness. A second is Bifurcated Architecture, which primarily addresses the problem of preserving the integrity of the controls while reducing maintenance costs over the lifecycle of the system (see pp. 155 and 157). The third extension, the Meta-Artifact, facilitates the first two, emphasizing the interaction among all three. This informal research hypothesis is stated formally in the following section.

Formal Research Hypothesis

A systems development process, augmenting OODA, with the following capabilities, provides a solution to the Constrained Problem:

1. A capability to comprehensively articulates stakeholders' explicit and tacit knowledge of a domain as artifacts and a means to store the artifacts as information that can be converted back to active knowledge, to assure adequate

understanding and proper implementation of control needs for the domain during the complete lifecycle of systems in the domain

2. A capability to convert the information representing the artifacts of the domain into active knowledge to assure adequate understanding and proper implementation of control needs for the domain during the complete lifecycle of systems in the domain

3. A capability to use the explicit knowledge of capability 2 to take account of existing systems and potential future systems in the domain during the complete lifecycle of the system under development, because of both the explicit and tacit knowledge contained in other systems on which the system under development may depend, especially when they must interact with each other

4. A capability to reduce or eliminate the gaps among development disciplines and development phases

5. A capability to produce an architecture that reduces the need to change software source code for maintenance resulting from breakdowns or changing control needs, with resulting reductions in costs and defects

The questionnaire-based and interview-based surveys for the evaluation (Chapter 6), discussed in the Validation section below, are formulated to assess the effectiveness of the Domain Rules Analysis extension in providing a solution to the constrained problem. Chapter 2 describes Domain Rules Analysis and the Meta-Artifact as the embodiments of capabilities 1 and 2, supported by appropriate enabling technologies (referred to as enablers; see Table 6 and Figure 6); the Meta-Artifact Process (MAP) as the embodiment of capabilities 3 and 4; and the Bifurcated Architecture as the embodiment of capability 5.

Validation

Validation will be familiar to the reader as a fundamental part of research. Two types of validation are used for the Meta Artifact Process (MAP): the prototype and the evaluation. The prototype is a "slice of life" example [222], making extensive use of integrated COTS tools and demonstrating all eight elements of MAP (see Chapter 2). To ensure that it is a realistic "slice of life," the prototype uses standard, robust graphical user interfaces and database management, as well as a formal UML model. The realism in the prototype is in the sense of having examples that illustrate all of the features of MAP, without necessarily having the scope of a full application. Nevertheless, because the examples use commercial grade tools for user interfaces, database management, and modeling, they could easily scale up to a full application by simply adding detail. The prototype also uses the Engineering Construction technique ([46]; see Research Approach section).

The prototype is described in detail to allow the reader to match concrete examples with each step (Chapter 3) of MAP (through annotated cross references from the examples to the steps) for applying the solution (MAP). Further, the prototype is executable, to allow the results of presented in *Breakdowns* to be reproduced. Finally, the prototype is a quasi-experiment [257] where both the process and the person (the author) applying the process are known in advance; neither the subjects nor the treatments are chosen at random (see Chapter 6, Future Research Directions section). Quasi-experiments "… are often used in software engineering when random samples of, for example, subjects (participants) are infeasible [257]."

The evaluation part of the validation [222] applies *Six Sigma* (see Appendix B; [184]) to the use of one of MAP's extensions [see Table 6], Domain Rules Analysis. Six sigma is a comprehensive and flexible system for process improvement. Six Sigma is not a case study, so case-study terms such as *explanatory, exploratory*, or *descriptive* do not fully apply [260]. However, the evaluation uses the Scientific-Interpretivist techniques (which includes the case study method) of field study, questionnaire-based survey, and interview-based survey [46]. To the extent of asking *what, how*, and *why* questions, the survey will be both explanatory and exploratory [236]. The evaluation has the following additional characteristics of a case study [207]:

- It illuminates a decision or set of decisions [260]
  - o Why they were taken
  - o How they were implemented
  - o With what result
- It investigates a contemporary phenomenon within its real-life context, especially when the boundaries between phenomenon and context are not clearly evident.

- It focuses on understanding the dynamics present within a single setting [73]

- It tests the theory the formal research hypothesis stated in the previous section

- It examines contemporary events

- It uses multiple tools and methods of data collection

- It uses direct observation

- It relies on multiple investigators

The six sigma evaluation is also a quasi-experiment [257]. The details for this six sigma evaluation are presented in the Results section of Chapter 6.

The prototype, described in the Prototype section below and in Chapter 4 and Chapter 5, provides a practical demonstration that MAP satisfies the conditions (provides the capabilities) of the formal research hypothesis (see Table 25), as stated above, including the articulating of tacit knowledge as described in conjunction with extracting domain rules in Chapter 3. Chapter 3 provides the detailed procedure for applying MAP that was used to build the prototype. Comparisons of MAP with other processes (Chapter 2, especially Table 9, and Chapter 6, Evaluation), successful demonstration and validation of MAP (the prototype shows that MAP does what is intended – Chapter 4, Chapter 5, and Chapter 6), and the review of existing literature (see Survey and Review of the Literature of Prior Research section), show that MAP is an improved systems development process.

Research Approach

Clarke provides a taxonomy of research approaches in [46]. Table 4 summarizes

Clarke's taxonomy. The Xs in Table 4 indicate which techniques and categories apply to

the three research traditions in Clarke's taxonomy; the asterisks in the first column

indicate which techniques and categories apply to the research in *Breakdowns*.

Table 4: Research Traditions and Categories

Technique or Category	Research Tradition		
	Engineering	Conventional Scientific	Interpretivist
*Non-empirical techniques	X	X	X
*Scientific techniques	X	X	
Interpretivist techniques	X		X
*Scientific/Interpretivist techniques	X	X	X
*Engineering: construction technique	X		
Engineering: destruction technique	X		
Pure research	X	X	
Instrumentalist:	X	X	X

Technique or Category	Research Tradition		
	Engineering	Conventional Scientific	Interpretivist
applied			
*Instrumentalist: problem-oriented	X	X	X
Underlying theory: descriptive	X	X	X
Underlying theory: explanatory	X	X	X
*Underlying theory: predictive	X	X	
Policy: prescriptive	X		
*Policy: normative	X		

While Scientific-Interpretivist techniques (field study, questionnaire-based survey, and interview-based survey) and a quasi-experiment were used in the validation, neither a full case study nor a full experiment was within the scope of *Breakdowns*. "The only real evaluation of a process or process improvement proposal is to have people using it, since the process is just a description until it is used by people. …empirical studies are not often used within computer science and software engineering [257]." Because of the need to involve highly skilled people in the application of a process (e.g., MAP) for a case study or an experiment, the cost of either undertaking is likely to be prohibitive. Nevertheless, case studies and experiments are potential areas of future research (see Chapter 6, Future Research Directions for consideration of case studies and experiments).

In the category selected for historical comparison of MAP to other methods, Non-Empirical techniques [46], the review of existing literature technique is provided (see Survey and Review of the Literature of Prior Research section). The Engineering: construction technique is used for both the development and validation of the MAP prototype. In keeping with the definition in [46], the construction technique was used for the conceptualization, design, creation, and demonstration of a prototype to apply MAP. According to the taxonomy of [222], a prototype is an example that validates MAP. The research is Instrumentalist: problem-oriented because it begins with a problem and prototypes a new process (MAP) in an effort to devise a solution [46]. The research is predictive [46] in that the underlying theory set forth for object-oriented domain analysis (see Domain Rules Analysis section in Chapter 2 for a detailed discussion of the underlying theory) predicts that frameworks and patterns (see Appendix B, Framework and Pattern) meeting the requirements of the domain will be constructed. The research has a policy dimension [46] because the enterprise would have to choose to perform domain-wide analysis, for example to support product lines.

Survey and Review of the Literature of Prior Research

General Problem

These works do not deal with the general problem directly, but they are not limited to the two domains of the constrained problem or to the internal controls targeted by the Prototype (see Prototype and Internal Controls sections below).

Extensive research has been done into the benefits of Object-Oriented Technology (OOT) for software reuse [19, 21, 22, 25, 43, 44, 45, 47, 48, 49, 56, 61, 84, 94, 149, 155, 167, 200, 200, 204, and 210]. Research has also demonstrated the value of Object-Oriented Domain Analysis (OODA) – applying OOT to the analysis of an entire problem domain, such as manufacturing or telecommunications [106 and 247]; see the Domain Rules Analysis section in Chapter 2 for a summary of domain analysis and OODA. A major obstacle to wider use of OODA has been the difficulty in defining a domain, leaving it a largely standalone activity that has not been integrated into a complete systems development process [49, 82, 168, and 202]

The research and application of OODA includes the following categories:

- Object-Oriented Domain Analysis [49, 82, 168, 200, and 235]
- Evolutionary Domain Life Cycle [94 and 96]
- Product-Line Development [43 and 67]
- Component-Based Development [32 and 56]
- Frameworks and Patterns (Appendix B; [49, 79, and 84]
- Architecture-Centric Methods [22, 149, and 169]

One approach in particular [235] is closely related to *Breakdowns* in its goal of extending OODA. The extension offered by [235] is based on the addition of technology requirements to the domain analysis, whereas my extensions are based on analytical and architectural concepts.

Processes for developing systems, where the processes are based on UML, with tool support and iterative, incremental methods, such as the Unified Software Development Process [22] and UML for Real-Time [160] address some of the process issues examined by *Breakdowns* (see Roles of Current Technologies and Extensions in Mitigating Gaps section), but leave out key enabling technologies of the proposed solution. For example, [22] does not incorporate automatic code generation and neither [22] nor [160] considers domain analysis for multiple systems; each focuses on one system at a time.

MAP is the only process that includes the three extensions (see Table 6). The Meta-Artifact, when applied through MAP, extends the modeling concept that the model is the application [160] by applying the concept to entire domains. The Meta-Artifact also extends the particulars of the concept in [160] beyond that of automatically generating code from the visual model (the basis for saying that the model is the application). That is, automatic code generation is just one sub-quality of a number of properties and qualities noted for the Meta-Artifact [see Meta-Artifact section].

Product-Line Development [43 and 67] and Component-Based Development [32 and 56] focus more on the management of core assets and components generated from them than on the process for producing the core assets or the nature of the core assets (such as the

Meta-Artifact); neither Product-Line Development nor Component-Based Development prescribes a particular process (such as MAP).

Two areas of research and application that are related to domain analysis that do not focus on OOT are Feature-Oriented Domain Analysis (FODA) and Enterprise Application Integration [31, 50, and 106].

Business-rule analysis [99 and 100] has been effective in looking at requirements from the perspective of business rules. Rule-based programming looks at creating entire software systems or components from business rules in the form of production rules [176]. A variation on creating entire systems or components from business rules (see p. 59) in the form of production rules is to focus on a particular aspect of a software application, then write that entire aspect in production rules [126]. One system uses a repository for business rule requirements elicitation and analysis [242]; another goes farther by using the repository of business rules to assist in the design process [209].

Certain aspects of expert-systems research are related to a portion of the proposed solution: business-rule-based analysis [99]; rule-based programming [176 and 237]; intelligent agents and multi-agent systems or architectures [78, 81, 159, 161, 179, 218, 219, and 252].

Constrained Problem

Work on related accounting theory and practice includes [9, 92, 130, 138, 142, and 205].

Related sources of research in AISs and descriptions of their application include [74, 89,

162, and 163]. See the Prototype section below for a summary of this literature related to

internal controls.

Works on Network-Centric Warfare include [3, 135, and 177]. Works on C2, C4ISR,

and Rules of Engagement include [54, 64, 135, 155, 238, 239, and 240]

Prototype

To narrow the scope for development of the proof of concept, the prototype (see Chapter

4, Chapter 5, and Chapter 6) further constrains the problem by including only one type of

control (internal controls, as defined below) within one of the two domains, the AIS

domain.

Kinds of Prototypes

The prototype is evolutionary rather than disposable. Disposable prototypes focus on

those aspects of the system that are visible to stakeholders, to assist in understanding the

requirements and validating their implementation. Shortcuts may be taken to rapidly

produce a prototype, e.g., a platform or language that is not suitable for the objective

system. Evolutionary prototypes based on OOT and iterative, incremental methods use

the language and platform of the objective-system. In addition to clarifying requirements

by demonstrating the visible aspects of the system (see Animated, rather than static, views of development artifacts section), an evolutionary prototype enables developers to produce early views of the objective system, then add detail to the prototype in successive iterations (see Iterative, Incremental Methods section), using the target platform and language [200 and 200].

As early views of the objective system, all aspects of the system and its architecture are included in an evolutionary prototype, at the level of detail of the current iteration. For example, drivers and stubs used for throwaway prototypes and structured integration testing (part of structured analysis in contrast to object-oriented analysis) are not necessary. Stubs do not allow any significant data to flow upward in the program structure, but the objects used in the evolutionary prototypes communicate with all other objects as they would in the objective system, within the level of detail for the iteration [200 and 200]. This is especially important for the early emergence of the architecture baseline, in that the detail excluded from early iterations, algorithms and data structures, is not an architectural concern. Thus, architectural frameworks and patterns (see Appendix B) for a domain can be produced without having all of the details for every potential target system. Therefore, the framework and patterns produced for the prototype can be applied to other target systems in the AIS domain. The prototype provides a framework, patterns, and representative components for software for the Purchasing Cycle of an Accounting Information System, focusing on the role of the extensions and enablers in MAP (see Table 6 and Figure 6).

Internal Controls

Internal controls are a major topic in accounting and audit literature [e.g., 51, 89, 92, 142, 205, and 229]. While internal controls are thoroughly defined there from various levels – the AIS itself is an internal control at the operational level [205, p. 109] along with planning and budgeting – and perspectives (objectives, procedures), the following definition, in the context of the accompanying definitions, is most appropriate for this research:

- Internal Controls: prescriptive statements whose purpose is to assure compliance with the business rules (see Table 20 and Table 21)

- Business Rules: declarative statements that define or constrain some aspect of the business (see Appendix B; [15, 22, 126, 99, 113, and 167]). In government and business, criteria for benefit eligibility or credit worthiness would be examples of business rules. In the military, Rules of Engagement would be examples. More generally, business rules represent *volatile variability* – those aspects of the system that change relatively often (see Business Rules section). References to business rules without explicitly specifying rules of engagement are intended to include rules of engagement.

- Domain Rules: rules based on the underlying principles, theory, longstanding practices, or traditions of the domain (Domain Rules section). Domain rules represent those aspects of a domain that seldom change, but they may drive business rules in that business rules must comply with domain rules.

Table 5 contrasts business rules with domain rules (also see related Table 11).

Table 5: Business vs. Domain Rules

Business Rules	Domain Rules
Short-term	Long-term
Volatile (changing over time)	Stable (unchanging over time)
Driven by current business context	Driven by application of a field of knowledge, such as accounting or command and control

The distinction between business rules and internal controls is also one of point of view and granularity – what is a business rule at one level could be an internal control at another level. For example, the following hierarchy starts with the business rule to safeguard assets, which is a general internal control objective:

- Safeguard assets
  - Require proper authorization of all transactions
    - Require separation of duties
      - Split procedures across two or more departments

- Edit employee department numbers to ensure that the transaction has been authorized by employees in different departments

Each successive level is an internal control on the preceding one. The types of internal controls that the prototype will be concerned with are at the level of the edit for employee department numbers. Software developers often consider any control implemented in software source code as an internal control. However, *Breakdowns* refers to internal controls in the narrower sense of only those controls that perform their prescriptive functions, as described above, from within application software, to distinguish them from software controls provided by the infrastructure, such as access controls provided by the operating system or software referential integrity controls provided by the data base management system (DBMS). Such infrastructure controls "...are mostly business-context-independent constraints [72, p.2]...."

As automation, integration, and network-centric operations have increased, the nature of transactions has changed so dramatically that entirely new approaches to internal controls have emerged, such as the use of artificial intelligence [9, 15, 23, 33, 34, 52, 74, 180, 223, 244, 245, 246]. For reasons such as those in Table 1, it is difficult to capture everything that was done by the manual, less automated, less integrated, or slower (e.g., traditional rather than e-commerce or network-centric applications) transactions.

Chapter 2 provides a detailed description of the Meta-Artifact Process (MAP) as a solution to the Constrained Problem. Chapter 2 describes the eight elements of MAP and their interdependence. The eight elements consist of three new concepts and five current technologies (the extensions and enablers referenced above). The enablers, as widely known technologies, are described only from the perspective of how they enable MAP.

Chapter 2: Meta-Artifact Process

Overview

This chapter describes in detail the Meta-Artifact Process (MAP) and its eight elements

(enabling technologies and extensions, Table 6 and Figure 6), which were described

briefly in the Solution Approach for the Constrained Problem (see Table 6). MAP is a

process to avoid breakdowns in meeting control needs in systems under the pragmatic

ethnographic-technological context and the theoretical knowledge management context

described in Chapter 1 (e.g., Problem Discussion and Solution Approach for the

Constrained Problem sections). The Meta-Artifact Process, applied through the

Procedure in Chapter 3 to develop the prototype described in Chapter 4, mitigates the

three persistent problems in software development – inadequate software reuse,

reliability, and interoperability (see Solution Approach for the Constrained Problem

section in Chapter 1 and Table 27 with related discussion) in a way that strengthens

controls throughout the system. Chapter 6 discusses how MAP satisfies the conditions

(provides the capabilities) of the Formal Research Hypothesis (see Table 25).

Table 6: Extensions and Enablers

Eight Elements of the Meta-Artifact Process (MAP)	
Extensions	**Enablers**
Meta-Artifact	Object-Oriented Technology (OOT)
Domain Rules Analysis	Visual Modeling with the Unified Modeling Language (UML)
Bifurcated Architecture	Integrated Modeling Tools
	Architecture Centricity
	Iterative, Incremental Methods

The next section describes the Meta-Artifact Process, with subsections on the Dynamic Interdependence of the Eight Elements of MAP and the Roles of Current Technologies and Extensions in Mitigating Gaps. The Solution Approach for the Constrained Problem section introduced the role of semantic and temporal gaps as causes of breakdowns. The subsection on the Roles of Current Technologies and Extensions in Mitigating Gaps summarizes conventional ways of mitigating such gaps and how MAP further mitigates the gaps (the nature of the gaps is elaborated in the section Contrasting the Meta-Artifact with Artifacts of Other Processes and shown graphically in Figure 12).

The discussion of the Meta-Artifact (one of the three extensions in Table 6) follows the section on MAP. A full explanation of the Meta-Artifact requires contrasting it with other artifacts of development. Contrasting the Meta-Artifact with other artifacts is intertwined with the processes, which produce those artifacts, and the process, which produces the Meta-Artifact, MAP, so other processes (and further aspects of MAP) are discussed in conjunction with their respective artifacts in Contrasting the Meta-Artifact with Artifacts of Other Processes. . Semantic and temporal gaps (see Contrasting the Meta-Artifact with Artifacts of Other Processes section) are the primary causes.

 Beginning with Domain Rules Analysis, the next seven major sections describe the remaining two extensions and the five enablers. The five enablers are discussed only to the extent of highlighting their interaction with the extensions, since details are readily available in the literature cited. A section on the Impact of MAP precedes the Summary at the end of this chapter.

Meta-Artifact Process

MAP is a process for developing software intensive systems that adds three extensions to *Object Oriented Domain Analysis* (see Table 6; [49, 82, 168, 200, and 235]).

MAP is an instance of *Model Driven Development* (MDD), an approach for developing software intensive systems, based on visual representations of artifacts [123], using visual modeling languages such as the Unified Modeling Language (UML).

MAP produces the Meta-Artifact, containing artifacts for an entire domain, including patterns and components from which software intensive systems can be composed with a *Model Driven Architecture* (MDA). MDA, another instance of MDD [4], uses standards (including UML) developed by the Object Management Group [181].

Figure 6, representing the overall flow of the Meta-Artifact Process, shows the dynamic interdependence of the eight elements of MAP – the three extensions (labeled Ex1-Ex3) and five enablers (labeled En1-En5) of Table 6 (see Dynamic Interdependence of the

Eight Elements of MAP). MAP uses knowledge management to discover system

requirements that may be hidden as tacit knowledge (see Problem Discussion section), as

well as explicit requirements. In terms of Figure 6, knowledge management techniques

would be applied, as shown, during the requirements and analysis activities of each

iteration, in order to convert previously tacit knowledge into active knowledge for

transformation into passive knowledge (information) in the form of artifacts in the Meta-

Artifact. With the iterative, incremental development methods referenced in Table 6,

each iteration includes the full set of development activities (core workflows in [22]); see

Iterative, Incremental Methods section.

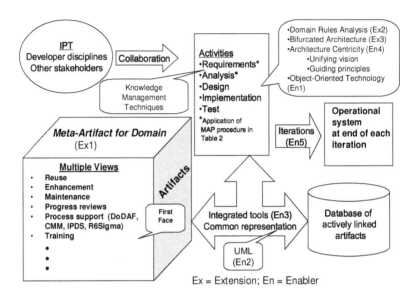

Figure 6: Dynamic Interdependence of the Eight Elements of MAP

Dynamic Interdependence of the Eight Elements of MAP

The interdependence of the eight elements is central to ensuring that the Meta-Artifact

provides the teleological remedy to the entropy in knowledge about the system that

otherwise occurs with time (see Meta-Artifact section). Domain Rules Analysis uses

domain rules to characterize and establish the boundaries of the domain and to increase

the completeness of the Meta-Artifact in representing the entire domain. In conjunction

with Domain Rules Analysis, the Bifurcated Architecture separates volatile variability (those aspects of the system that change relatively often) from *commonality* and *stable variability* (those aspects of the system that seldom change; see Domain Rules).

Architecture centricity adheres to the architecture that is embodied in the Meta-Artifact. Iterative, incremental methods assure that the Meta-Artifact represents the totality of the solution space (see Impact of Domain Rules Analysis section) from the outset, from requirements through executable artifacts (the increments). Object-Oriented Technology (OOT) provides by its nature certain capabilities (see Object-Oriented Technology section) that enhance control – inheritance for reuse and reliability, encapsulation for security and interoperability, and polymorphism for abstraction and extensibility. UML provides a formal graphical notation for OOT, which results in a common representation (see Unifying section below) for all of the artifacts of systems development. Integrated modeling tools leverage this common representation to capture, manipulate, and manage the artifacts of development for a system, from the highest-level semantics of broadly stated functional and non-functional requirements through operating or executing components. Together, the five enablers provide the technology infrastructure for producing and managing the Meta-Artifact, which is described next.

Roles of Current Technologies and Extensions in Mitigating Gaps

Developing systems with the Unified Modeling Language (UML), iterative, incremental methods, and visual models for MDD linked to running systems, mitigate the gaps described above.

- UML provides a common representation to mitigate semantic gaps across disciplines, workflow activities, iterations, phases, and individual systems

- Iterative, incremental methods (see Iterative, Incremental Methods section) mitigate both semantic and temporal gaps by performing the complete set of work flow activities (e.g., requirements, analysis, design, and implementation in the USDP) within an iteration in a relatively short time, increasing the availability of current artifacts and encouraging close collaboration among all disciplines at all levels of detail (i.e., requirements through implementation). This mitigation is weakened by the time that elapses between iterations and the even greater amount of time that elapses between phases (groups of iterations in USDP) and systems; the availability of current artifacts and their developers, for reference or collaboration, decreases as the elapsed time increases.

Two additional considerations increase the likelihood of semantic gaps between systems. First, while a common representation may be used for all of the systems, they are likely to have different architectures with different semantics, to the extent that artifacts and those who created them for previous systems are not employed to produce subsequent systems. Second, the artifacts representing the different systems are likely to reside in different repositories, which interferes with access (see section Meta-Artifact regarding accessibility of artifacts) to the systems' artifacts. Both of these considerations hinder development of subsequent systems in terms of reuse and interoperability.

- The creation of visual models for MDD that are linked to a running system, through automatic code generation or other means such as the transformation process in MDA [181], would increase the likelihood of retaining accessible artifacts that accurately represent the running system, reducing semantic gaps and the impact of temporal gaps on developing that system. However, the focus on one system at a time (see section Survey and Review of the Literature of Prior Research) continues to cause semantic gaps compounded by temporal gaps.

MAP, as summarized graphically in Figure 6, uses the mitigation technologies (UML, iterative, incremental methods, and visual models linked to running systems) to produce the Meta-Artifact (see section Meta-Artifact), guided by the Meta-Artifact (see recursive use below) to strengthen the mitigation, provided by the mitigation technologies alone, of semantic gaps and temporal, especially those among systems, in the following ways:

- MAP ensures that knowledge about the systems in a domain (the Meta-Artifact contains artifacts for all systems in a domain, but subsequent references in this context will frequently use "system," rather than "systems in a domain," for brevity) is captured in the Meta-Artifact as passive knowledge in the form of artifacts (e.g., narratives and diagrams) that preserve the semantics of a system across iterations, phases, and systems
- MAP uses the Meta-Artifact recursively to guide the development of new artifacts, which automatically extend the Meta-Artifact as they are added to the repository (see section Meta-Artifact for discussion of placing artifacts in the

repository); that is, MAP uses the Meta-Artifact to produce results that extend the Meta-Artifact. With the integrated modeling tools, MAP converts the information represented in artifacts of previous development into active knowledge on which to base the development and generation of new artifacts that are consistent with the baseline architecture captured in the Meta-Artifact (see next bullet). This conversion is temporally independent, preventing the knowledge about previous development from becoming tacit with the passage of time (see section Meta-Artifact regarding knowledge entropy). The new artifacts may be derived through specialization, based on decomposition or other reuse, or related to previous artifacts only through the baseline architecture.

- MAP's domain-wide focus produces the Meta-Artifact with artifacts for an entire domain, including patterns and components from which target systems can be composed (see Chapter 4: Prototype Description), before completing generation of any target systems. MAP uses these domain-wide artifacts to capture the baseline architecture for a domain in the Meta-Artifact, during the early iterations of development (see Chapter 4: Prototype Description). All systems developed from the Meta-Artifact for a domain incorporate the common baseline architecture (see Chapter 4: Prototype Description); MAP applies domain-wide architecture centricity.

The early development of a domain-wide architecture and artifacts ensures that high-level semantic artifacts for new target systems, in addition to more detailed artifacts from any previous target systems, are available in the Meta-Artifact.

Producing domain-wide artifacts during initial development avoids the impact of temporal gaps involving overall domain knowledge, especially from domain experts whose availability when subsequent systems were developed would be problematic. Retaining these domain-wide artifacts in the Meta-Artifact in a single repository (see next bullet) makes their transformation into active knowledge through the integrated modeling tools temporally independent.

- Using a single repository for all artifacts for a domain (see section Meta-Artifact for discussion of placing artifacts in the repository), managed by integrated modeling tools, avoids temporal gaps related to identifying and accessing needed artifacts during subsequent development for the domain

- MAP also deals with the practical issues of keeping the same team members available during the time needed to develop multiple systems by allowing them to stay productive without moving to an unrelated project. Traditional systems development processes focus on one system at a time (see Survey and Review of the Literature of Prior Research section), increasing the likelihood of the intersystem gaps noted above.

Because of MAP's domain-wide focus, different disciplines can work in parallel on artifacts for a domain and for multiple systems. Domain experts can proceed with aspects of a domain that do not affect the first system, while developers begin work on the first system. When team members for a discipline finish their work on an iteration for the first system, they can move to the next system, then return to the first system, increasing the likelihood that at least a core team for each discipline could work on all phases of a system. To the extent that such parallel activities can be sustained using at least the same core members for each discipline, the team members can continue increasing and applying their expertise, further reducing semantic gaps. As an observation that raises issues outside the scope of this article, organizations would need to consider changes in their approval and budgeting patterns, as well as their project management practices, to take full advantage of domain-wide parallelism.

Within an iteration, each functional discipline works through each activity with little or no specialization between activities, using the same notation to reduce or eliminate semantic gaps among disciplines (fragmented views shown in Figure 12; see section Contrasting the Meta-Artifact with Artifacts of Other Processes for detailed discussion of the gaps shown in Figure 12). .

> The workers do not hand off work to each other in serial fashion. Instead, they work together throughout the effort, evolving levels of detail to address their areas of concern. One of the chief goals -- and challenges -- of an engineering process and an architecture framework is to provide a means for the various development stakeholders to communicate and align their design decisions [38].

The artifacts produced for each activity for each discipline are automatically placed in the database by the tools (Figure 6, Database of Actively linked Artifacts). The tools can then access the artifacts in the database to produce the views of the Meta-Artifact as needed, reconverting the information captured in the artifacts – including tacit knowledge – into active knowledge.

Meta-Artifact

The Meta-Artifact is the electronically linked (through hot links directly on the artifacts or a browser in the integrated modeling tools) set of all of the artifacts (natural language narratives, graphical representations, or software code in various forms – including any linked, executable code for a system – describing the desired and/or current systems) of development for all applications (systems) for a domain. The electronic linking of all of the artifacts of development contributes to the Meta-Artifact's representation of the total solution space (see Impact of Domain Rules Analysis section). The Meta-Artifact is amplified by the enabling technologies and extensions (see Table 6 and Figure 6) and the Meta-Artifact's recursive use in MAP for its own development (see Recursive property below).

The Meta-Artifact is the cube (Ex1) in the lower left-hand corner of Figure 6, consisting of all of the electronically linked artifacts of the system. The properties of the Meta-Artifact (see Properties section) depend on the dynamic interplay of the eight elements shown in Figure 6 and described in the Dynamic Interdependence of the Eight Elements of MAP section. Many of the Meta-Artifact's qualities depend in particular on the integrated modeling tools (En3; see Table 8). The displayed face of the cube in Figure 6 shows examples of the multiple views, provided by the Meta-Artifact, of the architecture (see Architecture Centricity section) the Meta-Artifact embodies.

Paralleling these multiple views of the architecture, the Meta-Artifact itself has multiple dimensions – the Multiple Views, the Artifacts, and Time, as shown by the cubes in

Figure 6, Figure 7, and Figure 8. That is, set M = the Meta-Artifact, set i = a view, j = an artifact, and t = time. Then M(i, j, t) would be view i of artifact j at time t.

As can be seen from the examples in Figure 7, the artifacts of development are not limited to UML primitives, such as an object or association. They are more likely to be complex artifacts consisting of many such primitives. Nor are the artifacts of the Meta-Artifact limited to object-oriented representations or UML notation. Natural language documents such as RFPs and less formal requests are included. Because the artifacts include the software code, the running systems of the domain (system for brevity) are also part of the Meta-Artifact.

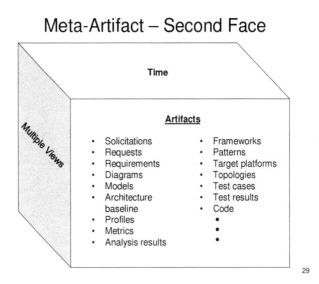

Figure 7: Artifacts in the Meta-Artifact

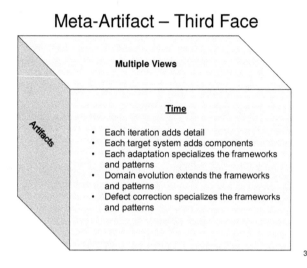

Figure 8: Meta-Artifact Over Time

The Meta-Artifact provides a knowledge management narrative for all systems in the domain that makes the background (tacit knowledge represented by a system) explicit. Narratives in knowledge management serve as a basic organizing principle of human cognition.

> Narratives, articulated as texts, can be seen as material traces of learning and collective remembering processes, social imprints of a meaningful course of events, documents and records of human action. They allow people to articulate knowledge through discourse [188].

Through their electronic linking in the Meta-Artifact, all artifacts, including software code, are part of the knowledge management narrative of the system and the domain. The Meta-Artifact, through its application in MAP, continuously supplies the teleological (purposely developed) remedy, through the active semantic chain (see below), to the entropy (loss of structure and explicitness) in knowledge about the system that otherwise occurs with time – as development progresses and during operation and maintenance of the system. That is, through the recursive application of the Meta-Artifact in the Meta-Artifact Process (MAP), knowledge captured in the Meta-Artifact can be converted into action and action into additional knowledge in the Meta-Artifact. The Meta-Artifact provides the structure and explicit representation of the knowledge about the system to prevent the entropy – loss of explicitness and structure – that would otherwise occur, by converting the tacit and explicit knowledge about the system into artifacts. The Meta-Artifact thus preserves the tacit and explicit knowledge of the system as information (in the form of artifacts) that can be converted into active knowledge, preventing the knowledge, through the use of the Meta-Artifact in MAP, from receding into history (becoming forgotten with the passage of time) and becoming tacit, including knowledge that had been tacit (e.g., taken for granted as a routine) before conversion into artifacts (see Background of the General Problem section). The Meta-Artifact preserves the ontology of the problem space (requirements for the domain) and solution space (systems satisfying the requirements – see Appendix B, Problem Space and Solution Space).

When breakdowns occur (see Appendix B, Breakdowns), stakeholders use the integrated modeling tools to create a view of the Meta-Artifact (IEEE std. 1471) to obtain understanding (see Active semantic chain section) of the background for the explicit knowledge they have and of the routines they perform against the background.

The Meta-Artifact eliminates the gap between the problem space and the solution space. Rather than being disconnected, as generally described in the literature [e.g., 49 and 230], they are different views of the Meta-Artifact. This at first seem would to contradict the problem-space focus of Domain Rules Analysis. However, because of the initial comprehensive focus on the problem space, the frameworks and patterns of the Meta-Artifact (viewed as part of the solution space) would be derived independently of the solution space (see Figure 17). The Meta-Artifact, when applied through MAP, extends the modeling concept that the model is the application by applying the concept to entire domains. The Meta-Artifact also extends the particulars of the concept beyond that of automatically generating code from the visual model (the basis for saying that the model is the application). That is, automatic code generation is just one of the sub-qualities noted for the Meta-Artifact (see Properties and Qualities sections below).

The electronic linking of the artifacts in the Meta-Artifact is comprehensive in the sense that the source code for software links all of the statements required to generate the executable code. Just as it is possible to corrupt the source code in some way that would break its linkage to the current executable code, it would be possible to break the global linkage within the Meta-Artifact. However, *Breakdowns* assumes the use of the tools at least as powerful as those used for the prototype (see Chapter 4, Chapter 5 and the section below, Integrated Modeling Tools), including tools for requirements and configuration management, that with reasonable care would ensure that the artifacts of the Meta-Artifact were synchronized from the highest-level semantics through the running system.

The procedure described in Chapter 3 for applying MAP includes steps to ensure the traceability among the artifacts, through successive iterations.

The Meta-Artifact performs a variety of actions as described in the following two sections, Properties and Qualities.

Properties

- Current

- Dynamic

- Prescriptive

- Unifying

- Seen from multiple views with a common representation (see Figure 6, Figure 7, Figure 8, and Figure 19)

- Recursive

- *Current*

The Meta-Artifact includes the running system and the artifacts from which it is generated. This allows MAP to use the Meta-Artifact to automatically generate the software code for the running system from the electronically linked artifacts, directly from certain graphical representations and other code, which in turn are electronically linked to the other artifacts. The artifacts are changed in order to generate changes in the running system, so they always reflect the current running system. This is related to qualities such as stakeholder access to views of interest (see Stakeholder access to current views appropriate to their interests section).

- *Dynamic*

A stakeholder's view of the system is not a single, static representation of a portion of the system or even the entire system that might have been prepared especially for the stakeholder (see Figure 11). The stakeholder views are based on the actual system and can be changed in realtime, as the stakeholder watches. The difference would be analogous to that between a sketch of the facade of a building prepared for an investor (stakeholder) during the early planning stages and the investor's standing at a selected location and looking at the facade of the completed building, then moving around and into the building, changing the investor's views in realtime. The building in this analogy would correspond to the aspects of the Meta-Artifact that allow stakeholders to have different views of the current running system, and to change those views in realtime. The building, however, would not have other aspects of the Meta-Artifact that would correspond to such things as blueprints for the building or the goods and services that would be produced within it. The Meta-Artifact includes the detailed specifications (blueprints) for building the running system, as well as the running system, which in turn can produce any of its outputs. This property is also related to the quality of providing stakeholder access to views of interest.

- *Prescriptive*

The Meta-Artifact, through its use in MAP, is prescriptive as a direct result of the automatic code generation provided by the integrated modeling tools. Combining this property through the integrated modeling tools with iterative methods and using a common representation for all of the artifacts, developer groups can add detail to the preceding group's artifacts, rather than translating existing artifacts into their own representation, effectively eliminating the interdisciplinary (semantic) gaps (see Solution Approach for the Constrained Problem section, Figure 12, and [47, 48,200, and 200]). Instead of causing turbulence in the development process with different representations for multiple developer disciplines, the Meta-Artifact supports collaborative and concurrent development that allows the different disciplines and stakeholders to enhance each other's activities (see Figure 6).

- *Unifying*

Unifying refers to systems in the upper right quadrant of Figure 9, which is adapted from a paper describing a four-valued logic for modeling human communication [206]. The x-axis in Figure 9 represents increasing degrees of integration (paralleling the levels of interoperability in the *Department of Defense Architecture Framework Version 1.0*). The y-axis represents increasing use for developing systems, across time and developer disciplines, of a common representation for the artifacts of development, a common architecture (within and across systems in a domain), and access to the artifacts by all stakeholders.

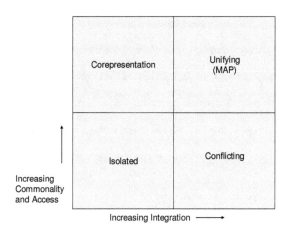

Figure 9: Unifying Role of the Meta-Artifact

In terms of the x-axis in Figure 9, systems in the Isolated quadrant have no interaction or interact with each other only through externally supplied data entered through displays or electronic files requiring manual intervention (e.g. converting and/or inserting the files in a drive bay). Systems interacting automatically through electronic exchanges, but without common user interfaces, shared databases, or direct interaction among software components (e.g. a client invoking services from a server) would be in the leftmost region of the Conflicting quadrant. Systems with the increasingly automated and direct interactions (common user interfaces, shared databases, and direct interaction among software components) would be in the rightmost region of the Conflicting quadrant.

The Conflicting quadrant in Figure 9 is so named because of the potential conflicts that may arise when interacting systems have disconnected artifacts and architectures (resulting, for example, from the use of different development methodologies and lack of a common representation for the artifacts) above the level of software code that are manifested in the code. These disconnects result in such problems as naming inconsistency (data and functions), inconsistency among function signatures (parameter lists), and data inconsistency (format, structure, and access methods).

In terms of the y-axis, movement upward begins to remove the disconnects among systems, by using a common representation for the artifacts of development, a common architecture, and access by stakeholders to the artifacts for the systems. As systems are moved rightward with increasing integration and upward with increasing commonality of representation for the artifacts of development, commonality of architecture, and access for stakeholders to the artifacts, systems move into the Unifying quadrant. In the Unifying quadrant, all systems use a common representation for artifacts above the level of software code and interact with at least common user interfaces, shared databases, or direct interaction among software components.

Systems that interact in all three ways (common user interfaces, shared databases, and direct interaction among software components), have a common architecture in addition to a common representation for their artifacts, and provide access to the artifacts for all of the interacting systems to all stakeholders for any of the systems would be in the upper right region of the Unifying quadrant.

Movement upward out of the Isolated quadrant, but with little or minimal integration would place systems in the Corepresentation quadrant (common representation and related accessibility). Individual applications produced with integrated modeling tools and the Unified Modeling Language (e.g., using the "the model is the application" process) would be in the Corepresentation quadrant.

The Elements, Properties and Qualities of the Meta-Artifact satisfy each of the conditions

Table 7: Unifying Quality of the Meta-Artifact

Conditions for Upper Right Region of Unifying Quadrant	Meta-Artifact Element, Property, or Quality
Common representation for artifacts above the level of software code	UML, Integrated Modeling Tools, OOT
Common user interfaces, shared databases, and direct interaction among software components	OOT, Domain Rules Analysis, Dynamic, Prescriptive, Animated views
Common architecture	Architecture centricity, Domain Rules Analysis, Dynamic, Recursive
Access to the artifacts for all of the interacting systems to all stakeholders for any of the systems	UML, Integrated Modeling Tools, Bifurcated Architecture, Current, Dynamic, Prescriptive, Multiple Views, Appropriate Views, Special Purpose Views, Active Semantic Chain

to place systems generated from it in the upper right region of the Unifying quadrant, as shown in Table 7.

- *Seen from multiple views with a common representation*

Multiple views with a common representation are provided by the Unified Modeling Language (UML – see Unified Modeling Language section) and the integrated modeling tools (see Integrated Modeling Tools). This property is essential to other properties and qualities such as the Current property and the quality Stakeholder access to current views appropriate to their interests.

* *Recursive*

With strong tool support, the Meta-Artifact embodies the system architecture (IEEE std. 1471) and becomes, as a management view, a database for all information about the system (e.g., see Database of Actively Linked Artifacts, Figure 6 and Database in Figure 11). The architecture embodied in the Meta-Artifact provides the unifying vision and guiding principles for the Meta-Artifact's structure (form) and behavior (function), so that the Meta-Artifact relates to the set of all artifacts of the system as a Metamodel relates to the concepts of a domain [166]: the Meta-Artifact provides the rules and structure for its own continued development (see Role of the Meta-Artifact in MAP section).

Consistent with the Unified Software Development Process [22], stakeholders do not start MAP with a preconceived architecture. Rather, the architecture emerges from the application of MAP. Architecture is part of the solution and MAP, through Domain Rules Analysis (see Domain Rules Analysis section), emphasizes thoroughly analyzing the problem space before moving to the solution space. Deriving the architecture through MAP allows the Meta-Artifact to recursively guide the process of developing the architecture (and the rest of the domain analysis), while also embodying the architecture. MAP thus overcomes a common weakness of OODA in moving prematurely to the solution space (see Domain Rules Analysis section below) by using the Meta-Artifact to derive the architecture from the requirements.

The unifying vision provided by the architecture contributes to the unifying property of the Meta-Artifact (see Figure 9; Unifying section). This unifying aspect of the Meta-Artifact, facilitated by the integrated modeling tools and UML, assures coherent semantics throughout the Meta-Artifact – from individual elements of simple artifacts through complex artifacts of structures and collaborations – again paralleling a metamodel [166]. This semantic coherence is important because the common syntax, naming consistency, and common representation enforced by UML-based tools are not sufficient to provide coherent semantics throughout the system – without the architecture, different parts and views of the system might seem disconnected. E.g., the use of repeated patterns in different parts of the system serves to unify it by greatly reducing the cognitive difficulty in understanding what those parts do [254]. This would avoid the situation shown in Figure 10, where both component A and component B provide the same service, but with different patterns. Developers and other stakeholders would need to interpret two different patterns for the same service, doubling the effort that would be required if the same pattern were used by both components.

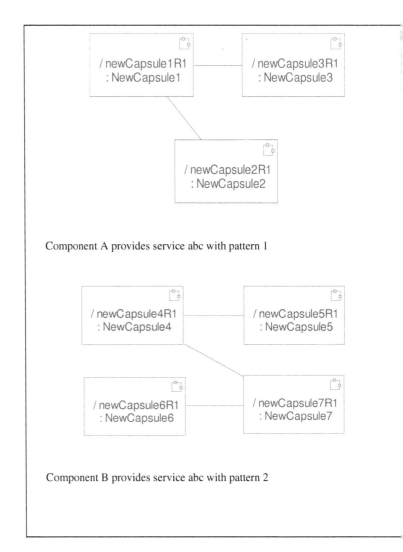

Component A provides service abc with pattern 1

Component B provides service abc with pattern 2

Figure 10: Different patterns providing same service

Steps 14 and 15 of the Procedure for applying MAP (see Chapter 3) include the

identification of repeated patterns. An extension of UML for formal methods would

further strengthen the semantic coherence of the model, as discussed in Chapter 6, User-

Friendly Access to Formal Methods.

As noted, the properties of the Meta-Artifact depend on the interplay of the eight

elements described in Table 6 (see Dynamic Interdependence of the Eight Elements of

MAP section). In particular, the nature and properties of the Meta-Artifact are realized

through two capabilities provided by the integrated modeling tools, as shown in Table 8.

Table 8: Role Played by Integrated modeling tools in Creating the Meta-Artifact

Capabilities Provided by Integrated modeling tools	Properties				
	Current	Dynamic	Prescriptive	Integrated	Multiple Views
An active link among the system's development artifacts	X	X	X	X	X
Automatic code generation		X	X		

Qualities

The above properties support the following qualities of the Meta-Artifact:

- *Stakeholder access to current views appropriate to their interests*
    - o Current because they are views of the same Meta-Artifact of which the running system is also a view, or as Andrew Lyons noted "...the model is the application [160]."
    - o Appropriate because stakeholders can use the integrated modeling tools to navigate through the Meta-Artifact to views that serve their current purpose

This quality of the Meta-Artifact strongly supports the IEEE standard for the architectural description of software intensive systems (IEEE std 1471), which states that multiple views of a single architecture should be provided, rather than separate architectures, for the many stakeholders that may be involved in a large system. Because the Meta-Artifact resides in a database (e.g., see Figure 6, Database of Actively Linked Artifacts) with convenient viewing tools, the stakeholders, including those from multiple engineering disciplines (e.g., network, hardware, software, human factors, or database) can use the Meta-Artifact, as the embodiment of the system architecture, for their respective purposes. Stakeholders can also produce special-purpose views not directly required to build the system, such as the Systems, Functional, Technical, and Operational views for Department of Defense Architecture Framework (DoDAF – see the Special-purpose views section and Figure 11).

Maintaining a single system architecture (see Artifacts of Other Processes in Practice and Architecture Centricity sections) through the Meta-Artifact from which other views can be derived with a common representation (UML, as discussed below) eliminates the need to reconcile separate architectures at integration or build time and avoids the semantic and conceptual breakdowns among developer disciplines (see Solution Approach for the Constrained Problem section above). All stakeholders can have access to the same views and development artifacts, allowing them to effectively collaborate and avoid not taking account of each other's requirements by making the tacit knowledge of each discipline explicit and the explicit knowledge of the disciplines understandable to each other.

UML provides the common representation for the Meta-Artifact (see the Unified Modeling Language section). The integrated modeling tools (see the Integrated Modeling Tools section) give stakeholders access to views of the system familiar to their differing mental models [254].

Unless special actions are taken to restrict stakeholders to viewing only certain artifacts, the tools allow any stakeholder to browse through all of the artifacts. Examples of such restrictions would be those enforced by configuration management tools or the selected views produced by the integrated documentation tools to extract only specified parts of artifacts of interest. The limited views created by the documentation tools would be analogous to providing subschemas for restricted or constrained stakeholder database needs. The configuration management tools allow only authorized developers to change the artifacts, even within a restricted view, preserving the linkage among artifacts discussed above at the beginning of the Meta-Artifact section. The UML specification provides underlying support for multiple views, as summarized in Figure 19. By supporting such views and providing a common representation for understanding among the stakeholders, UML contributes to semantic consistency among the artifacts [47, 48, 200, and 200], which is augmented by the linkage that allows navigation among artifacts, providing semantic detail at the level desired by stakeholders.

- *Animated, rather than static, views of development artifacts*

  Once the stakeholders agree that the visual model accurately captures their requirements, the Meta-Artifact carries the consistent UML representation through to the code [47, 48, 200, and 200], by using tools that generate the software code from the model. Because the automatically-generated software code is one of the actively linked artifacts, stakeholders can observe not only the static model, but also the animated visual model as it is animated by the executing code, allowing stakeholders to literally see that the code is doing what they intended, providing visual verification and validation (VV&V) that it is doing the right thing right [1]. Further, the running code supports analysis to assess how well the system architecture meets non-functional performance requirements. Even the look and feel of the system can be assessed early, through stakeholder interaction with prototypes as they evolve across multiple iterations (see Iterative, Incremental Methods section). Such VV&V reinforces the global consistency imposed by the tools, as discussed above (see Meta-Artifact section) and in Tools below, in that stakeholders can see not only how the developers have interpreted their requirements, but also those of other stakeholders.

This VV&V capability is amplified by the iterative, incremental methods in MAP. That is, stakeholders can apply this technique from the earliest stages of development, to avoid costly mistakes later. The value of early agreement among the stakeholders on what the system is to do and whether it is doing it is well described in [153]. VV&V contributes to the unifying property of the Meta-Artifact by allowing stakeholders literally to see that the system is, or is not, doing what they wanted it to do. Through application of VV&V during successive iterations, VV&V helps assure consistency between stakeholder requirements and the system during the system's full lifecycle.

After a system is implemented, the visual model – always actively connected to the running code – can be used for adapting the system to changing needs and technology in the current domain or extending it for an evolving domain. Generating the code from the visual model, after it has passed VV&V contributes to reliability throughout the lifecycle of the target systems.

- *Special-purpose views*

    The actively linked, electronically stored artifacts that comprise the Meta-Artifact are available to provide documentation in various forms to support, for example, Integrated Product Development Systems (IPDSs), Capability Maturity Models (CMMs), or to produce any of the twenty-six products defined by the *DoD Architecture Framework Version 1.0*. [64]. Stakeholders can obtain such special-purpose views as byproducts, not directly required to produce the system, without requiring separate overhead functions or introducing delays.

Figure 11 shows this concept of byproducts graphically for the DoDAF [64] products. Producing the DoDAF products with Structured Analysis (see Artifacts of Other Processes in Practice section) methods without integrated tool support requires separate manual overhead activities [155]. The integrated modeling tools extract artifacts from the database in which the Meta-Artifact resides (see Figure 6, Database of Actively linked Artifacts), using the active links among the artifacts to automatically generate the products. Figure 22 shows a graphics-based product (OV-6c) generated with the main visual modeling tool. Table 12 and

Table 13 show examples of text-based products (AV-2) generated by a documentation tool integrated with the model element database (see Figure 11).

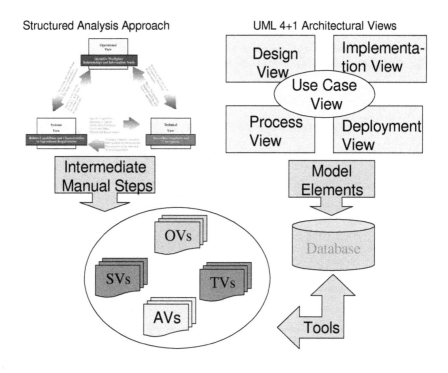

Figure 11: Special Purpose Views

Producing special-purpose views without separate overhead activities is not only a significant cost savings, but strongly supports the philosophy of letting developers be developers. Separate overhead activities and delays introduce direct costs and they also introduce the risk of indirect costs through artifacts that misrepresent the actual system

Letting developers be developers increases the likelihood that artifacts will be produced timely and properly. This is part of the new way of thinking about the system encouraged by MAP (see Contrasting the Meta-Artifact with Artifacts of Other Processes section). Because the Meta-Artifact, residing in its database (see Figure 11, Database of Actively linked Artifacts), is the system, work done on it via any of its views is not just a throwaway activity for communication purposes. Conversely, only those views that are an integral part of the system need be produced by developers; other views are byproducts generated through the tools. As byproducts, they do not take on a life of their own as often happens with large systems or organizations when development processes must be bureaucratically enforced.

If developers can benefit from a view, regardless of which stakeholder originated the view, as they produce the system (or the Meta-Artifact for MAP), with no need to stop to perform an overhead activity, they can produce the view without interrupting the process of building the system. If views that do not serve developers in this way, but that are needed by other stakeholders, can be prepared as byproducts, then the process of building the system is likewise not interrupted or weakened through resource or schedule contention. Electronic artifacts, created and managed by integrated modeling tools, make this possible.

- *Process and product are different views rather than separate things*

   Related to producing specialized views as byproducts, all activities performed by developers on the actively linked artifacts lead directly to the objective system, through the unifying and prescriptive properties. The executable increment generated for each iteration ensures that new artifacts in the Meta-Artifact are linked with previous ones, both through the compiling performed by the tools and the VV&V performed by the stakeholders. The compiling of the artifacts to generate an executable increment provides a comprehensive linking of the artifacts in the same sense that source code statements for software are comprehensively linked to the executable code generated from the source code statements (see p. 78). The comprehensive linking of artifacts in the Meta-Artifact, in turn, makes them consistent with each other in the manner that source code statements are consistent with each other.

   This unification of what has been in practice a troublesome disconnect [200 and 200] results from the automatic creation and linking of the elements of the artifacts by the tools, as MAP is applied and the Meta-Artifact is updated , and the storage of those artifacts in a database (see Figure 6, Database of Actively linked Artifacts) from which the tools can compose them into specialized products.

   - o Conversely, all changes to the objective system also change the actively linked artifacts (this is not necessary, nor is it preferred, as the objective system can be regenerated by first changing the higher-level artifacts of the Meta-Artifact).

- o From either direction, the active linking and integrated modeling tools preserve the current and unifying properties of the Meta-Artifact

- *Active semantic chain*

    What is referred to in *Breakdowns* as an *active semantic chain* for the system consists of the highest-level semantics of the system – requirements articulated as artifacts (e.g., requirements captured in use cases; see Figure 6 and Background of the General Problem section) – electronically connected through successive links in the chain – e.g., analysis and design artifacts such as class and collaboration diagrams – to the lowest-level semantics of the executable components (e.g., software code) and their outputs. It is created by the integrated modeling tools, linking the electronically stored artifacts and their elements (e.g., a class or association in a class diagram or a message in a collaboration diagram), and building on the semantics of the common representation provided by UML. Embodied in the Meta-Artifact, the active chain preserves the semantics of the system, from its genesis to its retirement. These semantics of the system are available as explicit knowledge, by application of the integrated modeling tools to the Meta-Artifact (e.g., by creating views appropriate to the stakeholder's current purpose). Access to electronically linked artifacts, in a single repository (Figure 6), for all systems in a domain avoids the documentation-related problems described in [60], which describes the difficulty in establishing and maintaining links for documentation in current practice.

Using the Meta-Artifact, the active semantic chain provides bi-directional narratives (see Background of the General Problem and Solution Approach for the Constrained Problem sections, and Appendix B, Narratives) – both to articulate what stakeholders know explicitly and to trace back to the rationale for what they know implicitly (experience-related knowledge or common sense that are tacit, or background, knowledge).

The active semantic chain is distinctly different from the relationship among artifacts found in practice. As discussed in Contrasting the Meta-Artifact with Artifacts of Other Processes, development activities in practice (e.g. requirements, analysis, design, and implementation) and their respective artifacts are physically and conceptually separate.

The active semantic chain is at the core of the Meta-Artifact's ability to promote understanding of the system. A key aspect of this understanding is documentation that is always current, readily accessible, and represented from the stakeholders' viewpoint (see [266] for impact of current documentation on semantics). Rather than being disconnected, as generally described in the literature [e.g., 49 and 200], the problem and solution spaces are different views of the Meta-Artifact (see Meta-Artifact section and Figure 6 and Figure 7), ensuring that documentation is always describes the current running system (see section Active semantic chain regarding always-current documentation). Such documentation has been the dream and the nightmare of systems development since the first question about how a system worked. The integrated modeling tools provide views tailored to the interests of stakeholders, based on all of the artifacts from which the current running system is derived, giving the stakeholders convenient access to documentation that is far more comprehensive than what likely to be available in current practice (see Contrasting the Meta-Artifact with Artifacts of Other Processes section).

A key byproduct of the active semantic chain is comprehensive traceability. Within a view, a stakeholder can navigate from any artifact to artifacts of higher- or lower-level semantics (or from courser to finer granularity) to assist in understanding, verifying, or validating the system. This traceability would also assist developers in gauging the impact of a change. The value of determining the impact of changes to the system is explained clearly in [19].

The importance of this active semantic chain and its many uses are described throughout this paper, beginning here with its foundations in the philosophy of science. Philosophers of science have written about the importance of capturing the genesis of ideas and the semantic chain of the artifacts that trace a development, whether a mathematical proof or a cultural tradition, from inception through to fruition. E.g., William Dilthey [69] concludes:

> The study of the concepts and precepts of existing cultural systems suffers from their not having the process that led to them '... preserved in its original fluid form' but rather 'objectified and compressed in the smallest possible form', i.e., in the shape of legal concepts.

Or as mathematician Jules Poincare observed in [192], we need to see "...the genesis [often a flash of intuitive insight] of our conceptions, in the proper sequence." That is, rather than retaining only the final, abstracted essence of the thought process, or in the case of mathematics, the minimum steps for the proof of the mathematical argument, we need to see how the thought process began and the intermediate reasoning that led to the abstracted essence. Poincare's words "proper sequence" capture the idea of how the thought process proceeds from imprecision, to precision over time. In software development, this would parallel the process of adding precision, through increasing detail, to requirements until the system is complete.

Impact on control needs – Because the active semantic chain of the Meta-Artifact improves understanding for development, operation, and maintenance (pp. 119, 155, 157, 170, and 187], it leads to fewer defects that result from inadequate knowledge of what the system is supposed to do, strengthening the integrity of controls (see p. 158, Table 27). Electronically linking the artifacts, including the informal statements of stakeholders, to formal artifacts the stakeholders can understand, then linking the formal artifacts to the running system through automatic code generation, reduces the misunderstanding that leads to defects. Such defects would include failure to capture requirements and failure to translate requirements into specifications that provide stakeholders with the system they need.

The failures to capture and translate requirements are worsened by tacit knowledge. The Meta-Artifact converts tacit knowledge into artifacts and preserves the knowledge of the system and embedded in the system (see [266] regarding undocumented semantics residing in software code and human consciousness) as explicit knowledge, accessible through the active semantic chain, preventing the knowledge from receding into history (becoming forgotten with the passage of time) and becoming tacit (see Appendix B, Tacit Knowledge). As discussed above (see Background of the General Problem section), if knowledge remains tacit it leads to breakdowns (see Appendix B, Breakdowns).

The Meta-Artifact enables derivation of frameworks and patterns that are stable over time. This temporal stability means that components generated from the frameworks and patterns do not require changes related to the requirements they initially implement, but only for adaptation or evolution of the domain (see Appendix B, Adaptability and Extensibility). Such temporal stability means that a wide range of target applications can be composed from the resulting stable components, plus new components generated from additional specialization of class hierarchies in frameworks ([49] e.g., see Figure 67 and Figure 68) and new patterns (see pp. 123). The temporal stability and increased reuse of frameworks, patterns, and components, in turn, preserves existing controls as target systems adapt and new target systems are composed (Table 27 and Chapter 4 and Chapter 5).

Contrasting the Meta-Artifact with Artifacts of Other Processes

Artifacts of Other Processes in Practice

Development of systems in practice involves a series of self-contained phases for the lifecycle of a system [47, 48, 200, 200, and 211]. Semantic and temporal gaps, which occur among phases and among developer disciplines (e.g., human factors, software, systems, database, network, or hardware engineers) within and across phases, hinder the ongoing development of a system because of the interdependencies among phases and among disciplines. Such gaps are magnified among systems that are developed at different times by different development teams, which may limit reuse of artifacts of systems development (artifacts; for examples, see Figure 6) and interoperability among the systems.

The use of phases in systems development processes provides structure and organization, to ensure that the requirements for a system to be developed are identified and analyzed so that the requirements can be met (implemented), resulting in a system that provides what the stakeholders want. The focus is on one system at a time (see section Survey and Review of the Literature of Prior Research). While a phase is largely completed before proceeding to the next, the phases are interdependent, with subsequent phases building on preceding ones, and with the likelihood of iteratively revisiting preceding phases (see Roles of Current Technologies and Extensions in Mitigating Gaps section and Figure 12).

Developer disciplines may use different representations– graphical notations, diagrams, or terminology – within and between the phases of a process (see subsection Roles of

Current Technologies and Extensions in Mitigating Gaps). For example, artifacts used in Structured Analysis (one of the most widely used processes [202, section 8.6] – such as functional decompositions, Entity Relationship Diagrams, Data Flow Diagrams [62], behavioral diagrams, and flowcharts [200] – have different representations. Further, different developer disciplines within a phase are likely to use entirely different methodologies as well as different representations. Software engineers might use an object oriented methodology while the database engineers might use a data-oriented methodology.

The use of different methodologies may also occur within a discipline, such as software engineering, for different phases – e.g., Structured Analysis through design, then object oriented coding. Finally, each discipline has its tacit knowledge (e.g., based on education and experience), shared by its practitioners, but not communicated to other disciplines. As with phases, the disciplines are interdependent, with various disciplines depending on each other to complete their work. The different representations, methodologies, and tacit knowledge of artifacts cause semantic gaps [47 and 48], which hinder the ongoing development of the system because of the interdependencies between phases and among disciplines. These semantic gaps are likely to be magnified between systems because the systems probably would be organized and managed independently, with different development teams. The impact of the intersystem semantic gaps is on interoperability and reuse (see subsection Roles of Current Technologies and Extensions in Mitigating Gaps).

Semantic gaps may be compounded by temporal gaps, which may occur for a variety of reasons between phases and systems, such as the following:

- Organizational budget and approval processes
- Staffing delays related to hiring or training
- Scheduling conflicts

Unless scheduling is closely synchronized among disciplines, there may also be temporal gaps from discipline to discipline within a phase, as each completes its work.

The impact on semantic gaps occurs when the temporal gap is sufficient that stakeholders (including subject matter experts and developers) and/or needed artifacts for a phase or discipline are not available for consultation or reference. A stakeholder who needs to understand the semantics of the system – e.g., to begin the next phase or related system, use, audit or maintain the operational system – may not have the means to do so efficiently and effectively. Temporal gaps are likely to be greatest between systems, where the influence of the organizational, staffing, and scheduling considerations noted above would be greatest. When the preceding artifacts are physically unavailable, too out of date to be of use, or simply not readily understood by other than the author, such artifacts recede to history (see Appendix B, Meta-Artifact and Time) and the information they represent becomes tacit. This is compounded when stakeholders think of the running system only as the software code (see p. 121) and do not attempt to use the other artifacts . Figure 12 highlights these semantic and temporal gaps graphically.

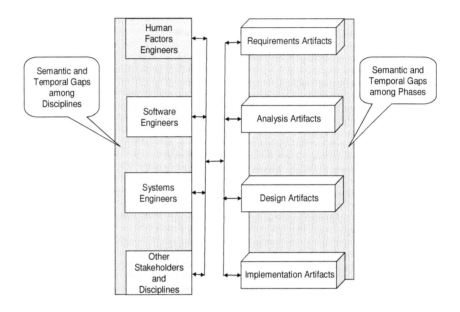

Figure 12: Current Practice

Far from providing an active semantic chain such as that in the Active semantic chain,

development artifacts produced for one phase may not be understandable to developers in

another phase – e.g., designers may prepare Entity Relationship Diagrams that coders

misinterpret because they are unfamiliar with the representation and semantics, or the coders may write programs the designers cannot read. The interaction among developers tends to be limited to the handoffs from one phase to the next. The result is that each developer or development group has a narrowly focused responsibility. Breakdowns occur when developers for a phase try to apply the knowledge and routines of their disciplines against a tacit background that differs from that of the preceding discipline (see Appendix B, Background). That is, the explicit knowledge as well as the tacit knowledge of the disciplines differs (see the Solution Approach for the Constrained Problem section above).

In practice, development projects do not comply with IEEE std 1471, which recommends having a single system architecture. Rather than separate architectures for each discipline, such as hardware and software, developers should have their respective views of the same system architecture, e.g., a hardware view or a software view. Separate architectures widen the semantic gaps among disciplines by adding differences based on architecture, such as different sets of guiding principles and different visions instead of a unifying vision for the system (see the Architecture Centricity section below). The breakdowns, including breakdowns in controls, caused by such architectural gaps are likely to involve substantial rework, in addition to breakdowns caused by the other disconnects independently of architectural gaps. The problem of disparate architectures is amplified when multiple systems in a domain must interact with each other, but are not based on a single domain architecture (see Domain Rules Analysis).

Coad and Yourdon describe the semantic gaps [254] within and among disciplines resulting from such representational, methodological, and architectural problems as disastrous over time [47 and 48]. With the passage of time, explicit knowledge articulated as narratives during the previous phase recedes into history (Appendix B, Meta-Artifact and Time) to become tacit knowledge – creating temporal gaps (see Figure 12) – unless converted into artifacts that can be converted back to explicit knowledge. Semantic gaps occur primarily between disciplines and temporal gaps between phases, as shown in Figure 12. Nevertheless, semantic gaps may also occur between phases for the same discipline because the representational, methodological, and architectural practices for a given discipline in one phase may differ from those of the same discipline in another phase. Likewise, if the work of disciplines within a phase is not carefully synchronized, there may be temporal gaps within a phase as one discipline waits for another.

The semantic and temporal gaps, or disconnects, often require a successive development group to start largely from scratch because they are unable to effectively use the artifacts prepared by the preceding group (interphase and/or interdisciplinary). During operations, when breakdowns are corrected or enhancements are required, these same gaps could be repeatedly encountered. The gaps are amplified when systems must interact with other systems in their respective domains. That is, the likelihood of representational, methodological, and architectural disparities increases with the likelihood that totally different development groups worked on the different systems.

Table 9 assesses the degree (low =1, medium = 2, high = 3) to which Structured Analysis

Methodologies ("…an amalgam [of methods] that has evolved over [more than] 20 years

[200]") and representative (those with the most complete sets of artifacts) Object-

Oriented Methodologies, including MAP, mitigate the gaps discussed in this section and

accomplish essential high-level functions of systems development.

Table 9: Comparison of the Meta-Artifact with Artifacts of Other Processes

Criteria	Structured Analysis Methodologies	Object-Oriented Methodologies			
	Various	MAP	OODA	Model is App	MDA
Systematic Analysis of Application Requirements	3	3	3	3	3
Systematic Analysis of Domain Requirements (Problem Space)	1	3	3	1	1
Identification of Needs (tacit requirements)	1	3	1	1	1
Reuse of Artifacts from an Application	2	3	1	3	3
Reuse of Artifacts for a Domain	1	3	2	1	1
Reduction of Gaps between Phases	1	3	1	3	3
Reduction of Gaps between Disciplines	1	3	1	3	3
Reduction of Gaps between Applications (interoperability)	1	3	2	2	2
Common Representation for All Stakeholders	1	3	1	3	3
Multi-View Architecture (IEEE std 1471)	1	3	1	3	3
Reduction of Maintenance and Enhancement Costs	1	3	1	2	2
Reduction of Maintenance and Enhancement Errors (reliability)	1	3	1	2	2
Training	3	1	1	2	2

Criteria	Structured Analysis Methodologies	Object-Oriented Methodologies			
	Various	MAP	OODA	Model is App	MDA
Total Score	**18**	**37**	**19**	**29**	**29**

The higher scores for MAP compared to other object-oriented methodologies (except for OODA) reflect MAP's focus on an entire domain, the use of knowledge management techniques for tacit knowledge (Identification of Needs), and the use of the Bifurcated Architecture (Reduction in Maintenance Costs and Errors). Compared to OODA, the other object-oriented methodologies score higher (except for domain related activities) because OODA is less complete, focusing on the high-level artifacts for a domain (those for requirements and analysis), leaving open the methodologies for design and subsequent development activities. Training is the criterion for which MAP receives its lowest score. As with all of the OOT methodologies, MAP is newer and less widely used than Structured Analysis. In addition, MAP adds a level of abstraction and complexity to all of the methodologies, by representing an entire domain in the Meta-Artifact.

Impact on control needs – The entropy in the explicit knowledge (see Appendix B, Meta-Artifact and Time) due to temporal and semantic gaps between phases increases the likelihood of breakdowns in meeting stakeholders' control needs in each phase (see the Solution Approach for the Constrained Problem and Meta-Artifact sections above). The longer a breakdown remains undetected, the more costly it becomes to repair. For example, if code testers discover a deficiency in the design, rework is infamously more expensive than if developers had discovered the defect during the design stage. Obviously, it is easier to relocate a bulkhead on the drawing of a ship than after the ship is built.

Intent of Artifacts of Other Processes

While the intent of existing processes in producing their artifacts has been to solve many of the problems identified in the preceding section, so far none has overcome them in practice, as discussed in that section. In particular, the existing processes have not overcome the breakdowns caused by the temporal gaps among artifacts produced in different development phases or the semantic gaps among artifacts produced by different developer disciplines within a phase (see Solution Approach for the Constrained Problem, Prescriptive, and Artifacts of Other Processes in Practice sections). Nor, by failing to consider the knowledge management issue of tacit knowledge, do they deal with the other causes of breakdowns discussed in the Background of the General Problem section above. That is, the artifacts produced by existing processes fail to avoid the breakdowns considered by both the general and the constrained problems.

Perhaps the closest to overcoming the breakdowns, in theory, are those based on UML, with tool support and iterative, incremental methods, such as the Unified Software Development Process [22] and UML for Real-Time [160]. However, [22] does not incorporate all of the enabling technologies (e.g., automatic code generation) and [160] does not consider domain analysis. Neither of them incorporates the three extensions of MAP (two that contribute to the Meta-Artifact and the Meta-Artifact itself), nor do

Change Impact [19], Early Requirements Agreement [153}, and Technology Extension to Domain Analysis [235].

More detailed comparisons of the artifacts of the Meta-Artifact produced by MAP with the artifacts of other processes are shown in Table 10 and discussed in the following pages or sections: OODA in the Domain Rules Analysis section below; Rule-Based and Rules-Centric processes in the Bifurcated Architecture to Encompass Commonality and Variability section below; Technology Extension to Domain Analysis on p. 54; Change Impact on p. 99; Early Requirements Agreement on p. 92; the Metasystem, UML for Real-Time, Zachman Framework; and DoD Architecture Framework (DoDAF) 1.0 are discussed in the next section, Other Processes Compared to MAP.

Other Processes Compared to MAP in Terms of Artifacts

The term Model Driven Architecture (MDA) is now popular. The newly adopted specification for the Unified Modeling Language, UML 2.0, is the foundation for MDA, which also can be supported by several other OMG standards [181]. The visual modeling community for software, including vendors such as IBM, continues to use the more generic term, Model Driven Development (MDD) [4 and 123] to refer to the architecture-centric visual modeling approach for software development that MDA encourages. Like MAP and the idea that the model is the application, MDA is dependent on powerful model compilers to generate code from visual models.

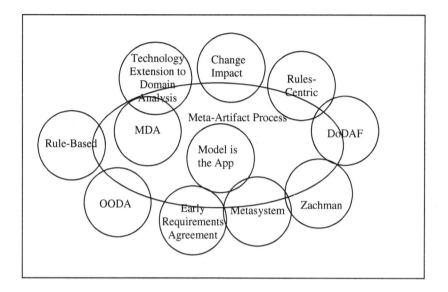

Figure 13: MAP Relationship to Other Processes

Figure 13 shows graphically the relationship of MAP to the processes identified in the

section Intent of Artifacts of Other Processes. Showing portions of some processes

outside of the oval for MAP indicates they perform certain functions extraneous to MAP,

such as detailed methods for defining a domain (rather than using domain rules; see

Domain Rules Analysis section below), structured analysis methods, or methods for

eliciting requirements. The portions within the oval for MAP are also performed by

MAP. The areas within the MAP oval, but outside the circles of the other processes,

indicate that artifacts of the Meta-Artifact produced by MAP reduce or eliminate the

problems identified in the section Artifacts of Other Processes in Practice, whereas the

artifacts of the other processes do not.

Table 10 compares the processes shown in Figure 13 in terms of the artifacts they produce compared to the properties, and qualities of the Meta-Artifact and related elements of MAP used to produce the Meta-Artifact. An "X" within a cell indicates that the property, quality, or element is supported by the artifacts produced by the process; a "—" indicates that it is not supported. A question mark in a cell indicates that the process is able in principle to support the property, quality, or element, but either does not do so in practice or, in the case of MDA, does so only if certain tools sets are selected. The property, quality or element is as defined above in its respective section.

Table 10: MAP vs. Other Processes

Property, Quality, or Element	Process or Methodology					
	MAP	OODA	MDA	Rule-Based	Model is the App	Technology Extension to Domain
Current	X	—	?	—	X	—
Dynamic	X	—		—	X	—
Prescriptive	X	—	X	—	X	—
Unified	X	—	?	—	X	—
Multiview	X	—	X	—	X	—
Stakeholder access to current views	X	—	?	—	X	—
Animated views of artifacts	X	—	?	—	X	—
Special-purpose views	X	—	?	—	X	—
Product = Process	X	—	—	—	?	—
Active semantic	X	—	—	—	?	—

Property, Quality, or Element	Process or Methodology					
	MAP	OODA	MDA	Rule-Based	Model is the App	Technology Extension to Domain
chain						
Domain Rules	X	–	–	–	–	–
Domain focus over entire lifecycle	X	?	–	–	–	?
Bifurcated Architecture	X	–	–	–	–	–
Automatic code generation	X	–	X	–	X	–
Integrated modeling tools	X	–	?	–	X	–

Property, Quality, or Element	Process or Methodology (continued)				
	Change Impact	Rules-Centric	Meta-system	Zachman	DoDAF
Current	–	–	–	–	–
Dynamic	–	–	–	–	–
Prescriptive	–	–	–	–	–
Unified	–	–	–	–	–
Multiview	–	–	–	–	X
Stakeholder access to current views	–	–	–	–	–
Animated views of artifacts	–	–	–	–	–
Special-purpose views	–	–	–	–	–
Product = Process	–	–	–	–	–
Active semantic chain	–	–	–	–	–
Domain Rules	–	–	–	–	–
Domain focus over entire lifecycle	?	–	–	–	–
Bifurcated Architecture	–	–	–	–	–
Automatic code genera-ton	–	–	–	–	–
Integrated modeling tools	–	–	–	–	–

The Zachman framework uses rows (roles in the design process, from context to lowest-level detail) and columns (products such as data and functions). The Zachman rows and columns are not electronically connected by an integrated tool set, using a common representation and methodology, leading to automatic code generation. Zachman did his work (80's to early 90's) before the widespread use of object-oriented technology, UML, or the tools that are now available. Even as generic as his framework is, it is data- (what) and function- (how) oriented, not object-oriented. Doing data-oriented analysis and functional analysis (traditional functional decomposition) - which together are referred to as Structured Analysis - results in disembodied data and functions. Zachman does not comply with IEEE std 1471 in having a single architecture with multiple views (the Zachman framework identifies seven architectures [261].

The DoD Architecture Framework (DoDAF) 1.0 suffers from the same underlying bias toward Structured Analysis as Zachman (although it does add some OOT concepts to the previous version). DoDAF does not define a process for developing a system, only documenting one. The products are disconnected in that one does not lead logically or intuitively to the next, as would be required in order to have a coherent process for developing a system, especially for automatic code generation.

While providing comprehensive documentation processes, Zachman and DoDAF with their related tools and approaches encourage disconnected development processes. This is due in large part to their being aggregations of techniques and tools, rather than an integrated tool set creating an active semantic chain using a common representation and methodology, leading to automatic code generation. The resulting gaps (see Solution

Approach for the Constrained Problem, Prescriptive, and Artifacts of Other Processes in

Practice sections) can be overcome only with extraordinary effort, involving overhead

activities that are not directly related to building the product (see the Process and product

are different views rather than separate things section).

Comparison of Other Domain Analysis and Visual Modeling Processes to MAP

The metasystem in [7] is the result of disconnected steps – domain analysis, domain

model, infrastructure specification, infrastructure implementation, and operational system

– analogous to those described above (see Solution Approach for the Constrained

Problem, Prescriptive, and Artifacts of Other Processes in Practice sections). The

metasystem is comprised of multiple repositories that are connected conceptually, but are

not electronically linked in such a way that successive ones are generated from previous

ones, using a common representation and methodology.

The idea that the model is the application [160] noted in the qualities of the Meta-Artifact

section, envisions generating code for a single application from a visual model, but does

not discuss consideration of an entire domain, with frameworks and patterns for reuse to

generate target systems. However, the artifacts leading to the generated code for the

single application are electronically linked to the code.

The Meta-Artifact offers the benefits of the metasystem for reuse and domain-wide

analysis, as well as the benefits offered by the model-is-the-application idea for

generating code from the visual model. MAP then provides a process to ensure that the

Meta-Artifact represents the totality of the solution space for the domain (see Impact of

Domain Rules Analysis section), accommodating both adaptability and extensibility

[Appendix B: Adaptability and Extensibility], by assuring that the Meta-Artifact is the

focus of system activities for its entire lifecycle. In this way, MAP combines the two

general aspects of domain modeling cited in [7]: the operational aspect and the

infrastructure aspect. Because these aspects are different views of the same electronically

linked artifacts, the gaps between the steps of the metasystem are also eliminated.

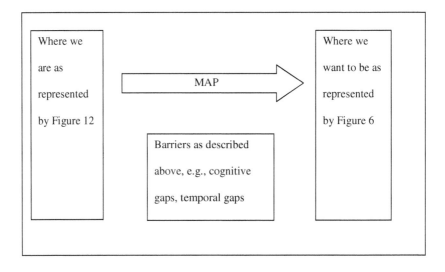

Figure 14: Overcoming Barriers

Figure 6 can also be contrasted in terms of obstacles, as in Figure 14. The list of artifacts

in Figure 7, while shown in the context of the Meta-Artifact, also provides typical examples of the individual artifacts that are produced in practice, which are static and declarative even if all are available and up to date. The static and declarative nature of the individual artifacts worsens the gaps described above (Solution Approach for the Constrained Problem, Prescriptive, and Artifacts of Other Processes in Practice sections).

In contrast, the Meta-Artifact is *dynamic*, *prescriptive*, and *unifying* (see Properties section) in that the electronic linking of artifacts expressed in a common representation seamlessly connects the artifacts [47, 48, 200, and 200]. With UML, these representations are visual or linked to visual representations.

Role of the Meta-Artifact in MAP

The role of the Meta-Artifact in MAP is defined by explicitly recognizing the Meta-Artifact for what it represents (as described in the Meta-Artifact, Contrasting the Meta-Artifact with Artifacts of Other Processes, and MAP sections above) and continuously applying it during the lifecycle of the system. In this way, the Meta-Artifact continuously supplies the teleological remedy to the entropy in knowledge about the system noted above (see Appendix B, Meta-Artifact and Time).

In terms of process, the departure from current practices is not achieved by eliminating or even reducing the details captured during development activities – e.g., thorough requirements gathering, analysis and design are no less essential, so in this respect, MAP does not depart from other methods. Even automatic code generation does not account for the nature of the difference between MAP and other methods. What changes is the role played in MAP by the results – captured in the Meta-Artifact (Figure 6) – of the development activities.

This role of the Meta-Artifact expands both the developer's view of the system and responsibility for the system. Because of the Meta-Artifact's properties, the results of development activities from successive phases no longer are of interest only during that phase, but become permanent, active links in the semantic chain, accessible to all developers (see p. 187). MAP depends on developers' ability to readily view the system beyond the artifacts they create to those they need to create them. When using MAP, developers are able to view and reason about the full system – the Meta-Artifact – in its current state, during all phases of development. Tacit knowledge embedded in the running system can be made explicit through the other artifacts of the system and accessed through the active semantic chain.

This implies that developers for a given discipline, e.g., software engineers, must learn the skills necessary for each phase, rather than specializing in how to build the artifacts of a single phase. In current practice, this specialization usually means learning the skills of one of the later phases, such as coding, becoming more proficient with those skills, then possibly advancing to the skills of earlier phases such as design or analysis. Organizationally, this separation of skills contributes to the gaps noted in the Contrasting the Meta-Artifact with Artifacts of Other Processes section. Another costly outcome of this current practice is the loss of hard-to-learn proficiency as developers advance from one set of skills to another. MAP turns this cost into a benefit by using the current practice for learning multiple sets of skills, but preserving developers' proficiency with a process that requires their continued use by each developer (perhaps with varying degrees of emphasis on large teams). Multi-skilled developers in turn reinforce the elimination, inherent in the Meta-Artifact, of gaps between development phases.

As the totality of the solution space (see Impact of Domain Rules Analysis section), with its role in MAP, the Meta-Artifact moves MAP still farther from other processes. The above changes in the roles and perspectives of individual developers are amplified through their application to the entire solution space during development, rather than a single system or application (see p. 182). For example, developers might consider the structure of frameworks early in the development process, to facilitate composition of target systems. Accordingly, stakeholder views, including those of developers, are of the entire solution space.

As additional systems or applications in the solution space are targeted for implementation, the Meta-Artifact provides views of all artifacts for systems already implemented, with higher-level artifacts – e.g., frameworks and patterns – for targets not yet implemented (see Figure 8). Developers, in turn have the skills and experience with the Meta-Artifact to reason about it efficiently and effectively. Ongoing use of MAP expands the Meta-Artifact, while the Meta-Artifact at the same time continues to recursively provide the rules and structure for its own expansion as it is used to implement additional target systems.

Gaps between development phases (see Solution Approach for the Constrained Problem, Prescriptive, and Artifacts of Other Processes in Practice sections) result in fragmented views of the system (see Figure 12). At the end of each phase, the artifacts produced are viewed as the system. E.g., when all of the requirements are officially collected, especially when there is a formal step in a contractual process that recognizes them as complete – possibly including an official acceptance ceremony – the set of documents recording the requirements is viewed as the system. When the system is in operation, the running system – the code – is thought of as "the system." Other artifacts, whether requirements-, analysis-, or design-related are thought of as being about the system, rather than the system. With MAP, the system is no longer thought of only in terms of the most recently completed artifacts, but in terms of many views of the Meta-Artifact.

This change in thinking applies during and after development, throughout the lifecycle of the system (see p. 119). It applies especially when stakeholders are implementing a new target system. That is, by thinking of the new system as creating another view of the Meta-Artifact, stakeholders are able to concentrate on the increment to the Meta-Artifact, using it as a guide and tool – through the integrated modeling toolset – to generate the increment, rather than having to create a complete set of development artifacts, guided by existing artifacts with the limitations described above (see Solution Approach for the Constrained Problem, Prescriptive, and Artifacts of Other Processes in Practice sections). A stakeholder can reason about the objective or operational system through the actively linked artifacts from which it is was automatically generated and composed, relying not on something about the system, but on a particular view of the operational system itself.

The Meta-Artifact also provides views of the problem space (see p. 156), for reference as new target systems are considered. As the domain evolves, the problem space view of the Meta-Artifact allows developers and stakeholders to consider only the increment due to evolution (see Figure 8).

The Meta-Artifact, as already noted, provides the rules and structure for its own subsequent development (see Appendix B, Meta-Artifact). That is, the Meta-Artifact is used recursively in the incremental building of itself by MAP, analogous to the metacircular role of the UML metamodel [19], except that the UML metamodel is the starting point of UML, whereas the Meta-Artifact is the product of MAP. The Meta-Artifact goes beyond the role of meta-models and meta-data, which provide the formalisms for defining other models or data, because it includes the completed artifacts themselves. MAP converts knowledge captured in the Meta-Artifact into action and action into additional knowledge in the Meta-Artifact.

Domain Rules Analysis

Domain Analysis Background

A domain, according to the *American Heritage Dictionary*, is a sphere of activity, concern, or function; a field. In systems engineering, domains may be vertical, such as a full range of applications for a particular type of business (e.g., including at least some aspects of every top level box under MIS in Figure 2: AIS Context) or weapons systems for the Department of Defense; or horizontal, such as accounting, office automation, middleware for distributed objects, or operating systems, providing support to many vertical domains. Combinations or subsets of these vertical and horizontal classifications are also commonly viewed as domains. Specific examples are: Accounting Information Systems (an example of combining horizontal and vertical domains by integrating a number of vertical applications in a way that can be used across multiple types of enterprises), Computer Aided Design, Computer Aided Manufacturing, Manufacturing, Command and Control systems, Combat Support, Modeling and Simulation, Weapon Systems, and Sensors (an example of a subset of a vertical domain, Weapon Systems).

Figure 15 shows the Department of Defense Joint Technical Architecture domains (JTA) [64] with their related subdomains reused across agencies, services, and commands through specialization and standard publish and subscribe interfaces enabled by the innate characteristics of OOT (see Object-Oriented Technology section) and integrated modeling tools. Domain Rules Analysis, a method for assuring a comprehensive analysis of the problem domain before developing a solution, helps produce a Meta-Artifact that represents the totality of the solution space (see Impact of Domain Rules Analysis section), including frameworks, patterns, and components from which target systems can be specialized and composed. Developers can then take full advantage of those characteristics of the Meta-Artifact to compose solutions for the domain and its subdomains, as described below and shown in Figure 17.

According to Prieto-Diaz [7], the term "domain analysis" was introduced in the context of software reuse by Neighbors, to mean "... the activity of identifying the objects and operations of a class of similar systems in a particular problem domain [175]." In theory, domain analysis focuses on the problem space, e.g., as Kang observes [140], domain analysis deals with "... the process of identifying, collecting, organizing, and representing the relevant information in a domain...." According to Cohen [49], this is done for the purpose of "... mapping from the problem space (the group of related systems and their essential requirements) to the solution space: a common architecture consisting of frameworks in a number of domains." Such frameworks

> ... embody an abstract design for a family of related problems in the form of a set of classes. ... By enforcing a focus on a group of systems rather than individual systems, domain analysis can support the creation of abstractions that will cover a range of applications. ... Domain analysis considers commonalities and differences across the group and provides a systematic approach for dealing with commonality. ... Developers can then use essential framework principles to fashion domain-specific design patterns for development of applications [49].

Pressman [200 and 200] notes that the end result of object-oriented domain analysis is the specification of classes that characterize the domain. Di Nitto [67] classifies groups of systems into product lines, product families, or families of product lines. Sessions [220] adds families of specifications, citing Java 2 Enterprise Edition (J2EE) as an example, while describing Microsoft® Visual Studio® .NET as a family of products.

Attempting to take account of a domain (problem space) in building systems (solution space) is nothing new, of course, especially in operating systems, going back at least as far as IBM's OS/360 line. Product lines for application software (solution space) have also been around for over thirty years in business application software, such as payroll systems and general ledger systems. Such software application product lines depended on parameter-driven feature choices when generating a target system, then parameter-driven options within the target system.

Current enterprise-wide systems such as SAP R/3 have combined this parameter-driven approach with extensive table-driven capabilities (e.g., charts of accounts, transaction codes, schedules, report options) to allow standard integrated systems to be tailored to an entire enterprise. Another approach hybridizes the multi-agent and parameter-table driven approach, embedding the production rules in collaborating agents, where the system is composed of agents selected and configured through parameters. In general, such approaches deal extremely well with commonality (see Appendix B, Commonality); they deal less well with variability (see Appendix B, Variability), because all of the software code must be written in advance. That is, all of the current and future capabilities for the desired range of applications must be anticipated and software code written to support them, so that they are available when specified by the parameters.

Software components that have not been generated for a domain from a common framework or set of frameworks for the domain can also be composed into target systems [57, 111, and 225], but they still face the same limitation as the parameter- and table-driven approaches in that all of the software code for a component must be developed in advance. In composable multi-agent systems, this would be mitigated to the extent that variability could be accommodated by learning on the part of agents. The appeal of such component-based development when legacy systems are involved, of course, is that the software code already exists – the challenge is the reverse engineering and development of software wrappers.

Frameworks have also been developed for applications within the AIS domain (see Figure 2), such as [63] and [129], but they do not deal directly with the breakdowns in meeting the control needs of systems where longstanding ethnographic conditions are confronted with evolving technology. (see the Discussion General section above). DIICOE [63] is an example of a framework for military applications, but it requires the development of software components with standard interfaces, rather than evolution of a framework of classes using inheritance and patterns [49 and 84].

Because of its inherent features, Object-Oriented Technology (OOT) for software allows a fuller realization of domain analysis (see pp. 123 and 140) than the above approaches, especially for variability. Unlike parameter- and table-driven techniques, OOT allows software frameworks (see Appendix B, Framework; [49 and 84]) to be developed that can accommodate variability without having to write all of the software code first, for whatever features and options are to be supported. The frameworks can be amplified with patterns to guide the composition of components and the implementation of software code within components, an approach used by Microsoft .NET and Sun J2EE [220].

In the two domains considered by *Breakdowns*, the combination of OOT and Domain Analysis – Object-Oriented Domain Analysis (OODA) – promises to overcome this limitation of developing all of the software code in advance, by producing software frameworks, patterns, and components generated from them for target systems [93 and

95]. There are other limitations to OODA that *Breakdowns* deals with through the extensions and enablers of MAP (see Table 6 and Figure 6), as noted above in this chapter, below in this section, and below in the section Bifurcated Architecture to Encompass Commonality and Variability. The section Bifurcated Architecture to Encompass Commonality and Variability discusses the use of two enhancements to the OODA approach to variability, one physical (see Figure 18) and one to distinguish stable from volatile variability (see Domain Rules section and Business Rules and Volatile Variability section). The bifurcation is within the software, between commonality and stable variability on the one hand and volatile variability on the other. (see Domain Rules section, Business Rules and Volatile Variability section, and Figure 18).

While the purpose of domain analysis is closely related to spending more extensive time in the problem space by looking broadly at an entire problem domain, domain analysis in practice [see p. 130] tends to move too quickly to the solution space. In [49], Cohen states that frameworks may evolve through development of several related systems. Pressman [200 and 200] lists five activities for the domain analyst to perform, but all five include existing systems. There are obvious reasons for this practice, including:

- Developers may have no clear means of establishing boundaries for the domain other than existing systems [202]
- This is a natural way to proceed, since the reason developers look at a domain is to build systems for it
- Existing systems may be the most practical source of requirements, through reverse engineering
- Existing systems may be reused

Such a focus on existing, planned, or preconceived solutions may limit the analysis of the domain in ways that defeat the purpose of domain analysis, e.g.:

- Little *a priori* consideration of future systems
- Systems that are not adaptable (see Appendix B, Adaptability) within the current domain
- Systems that are not extensible (see Appendix B, Extensibility) for domain evolution
- Systems with extensive options that may never be used, rather than frameworks and patterns for composing target systems tailored for selected requirements

There are several assumptions implicit in domain analysis theory combined with OOT:

- A domain can be identified
- There are groups of applications within a domain with more commonality than differences
- Frameworks can be built by mapping from the problem space to appropriate classes
- Design patterns can be derived from the frameworks
- Applications in the solution space can be composed from the frameworks and patterns

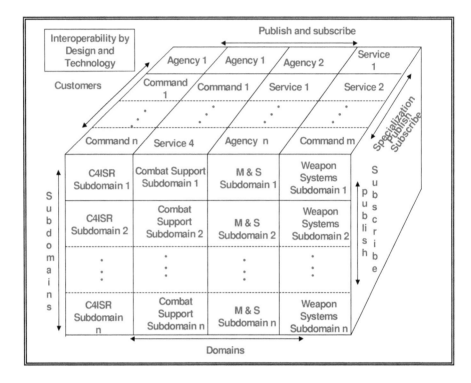

Figure 15: Domains

OOT intersects with Domain Rules Analysis by providing a structure for commonality

and variability through inheritance, encapsulation, and polymorphism. In this way, OOT

can be used to develop frameworks and patterns for the domain identified through

Domain Rules Analysis (see Definition of Domain Rules Analysis and Appendix B,

Framework and Pattern; [49 and 84]). Objects derived from the framework as components can be composed into target systems [93 and 95], e.g., for product lines [67], using integrated modeling tools, applying and identifying useful patterns.

Some integrated modeling tools (see Integrated Modeling Tools section) greatly expand this configuration process by generating code from the framework for interprocess communication, creating concurrent, distributed, real-time applications with publish and subscribe interfaces (see p. 256 for implementation discussion). Another feature of OOT, encapsulation, \ facilitates the interprocess communication by enabling interaction of components strictly based on services available through the publish-and-subscribe interfaces. Encapsulation also expands the use of patterns [84] both at a design and at an architectural level by allowing discrete services to be bundled for reuse in multiple patterns.

Even with the natural support that OOT gives domain analysis in principle, there is a practical contradiction. The goal of domain analysis is to consider an entire domain (the problem space), but as Cohen observes [49], object-oriented frameworks and patterns have tended to be connected to the solution space. This is an example of how domain analysis has relied in practice on defining the problem space from the solution space (see p. 127) e.g., extracting requirements from existing systems through reverse engineering. The dynamics of this practice and how to avoid it are discussed below, so that the frameworks and patterns are derived independently of the solution space (see Definition of Domain Rules Analysis section).

Coad [47] reinforces the need to concentrate on the problem space by noting that the most stable aspects of a system are the classes "...which strictly depict the problem domain." Because classes that characterize the domain are its most stable aspects, they should have a natural correspondence with the stable domain rules (see pp. 132 and 156). The stability of domain rules would also be consistent with the role of stability of problem domains in [7]. Domain Rules Analysis takes advantage of this natural correspondence between classes that characterize the domain and domain rules, as described below (see p. 156).

Impact on control needs – Failure to anticipate future capabilities by having comprehensive frameworks and patterns for a domain causes disconnects between the problem space and the solution space, with the potential for breakdowns in meeting control needs. Such disconnects and breakdowns would be handled as new features and options in subsequent versions and releases, but all of the software code for all of the new features and options still must be written in advance. Developers attempting to anticipate future capabilities must elicit requirements, whether related to analyzing trends (e.g., market growth, technology evolution) or anticipated control needs, adding the problems of not articulating tacit knowledge as artifacts for the less-certain future control needs and related trends to the risk of failure to articulate tacit knowledge of current control needs (see the Problem Discussion section).

Domain Rules

The ability to set the boundaries of a domain independently of existing systems and the solution space in general would be an essential step toward focusing on a thorough analysis of the problem space to produce frameworks likely to cover the entire domain [49, 94, 200, 200, 202, and 7]. A new concept, domain rules, provides boundaries for a domain in pure problem space terms, to keep the focus on the problem domain. Domain rules (see Table 16 and Table 17) are based on the underlying principles, theory, longstanding practices, or traditions of a domain, such as:

- Economic theory
- Laws of physics
- Principles of war
- Military doctrine
- Legal precedent
- Legislation
- Industry standards

Domain rules are used in conjunction with business rules (see Business Rules) to produce a Bifurcated Architecture (see Bifurcated Architecture to Encompass Commonality and Variability section).

The mechanism used by which domain rules identify the boundaries of a domain is to establish what must be in the domain (in the sense of completeness – see Procedure, steps

7 and 12) and what should be excluded. By this mechanism of inclusion or exclusion, domain rules serve to differentiate one domain from another and provide a reference baseline of the minimum requirements for the domain. E.g., does ensuring that debits and credits are in balance (Table 16) apply to C2Ss? Does placing the enemy in a position of disadvantage through the flexible application of combat power (Table 17) apply to AISs? Some domain rules might apply to multiple domains – of interest for reuse across multiple domains – e.g., prepare clear, uncomplicated plans and concise orders to ensure thorough understanding (Table 17). Such differentiation and the possibility of multiple domains with common domain rules would be useful in searching for reusable components from one domain to another. That is, applicable domain rules could be used to include components for consideration or exclude them from continued interest.

The minimum baseline of requirements provided by domain rules would serve as a corrective to the solution focus of a traditional request to develop a system (see Procedure, step 9) and supplemental sources of information (see Procedure, step 2). That is, capabilities that might not be inferred from the request or other sources would be revealed through assessing whether all domain rules were supported (see Procedure, steps 7 and 12). Such capabilities could suggest, for example, new functions, decompositions, or rearrangements of the traditional ones, not only to add robustness to the functions, but also to anticipate future needs. Conversely, requested capabilities that did not correspond to one or more domain rules could be challenged, either to be clarified or dropped. In principle, there is no reason this matching of domain rules and capabilities should not work in both directions, because a traditional function or process may have a domain rule embedded in it that should be made explicit (see Procedure, step 7), but the primary direction would be from the domain rules to the functions and processes.

Domain rules form the overarching category of requirements that characterize a domain by distinguishing it (and the systems that meet the needs for the domain) from other domains. The function of domain rules in characterizing the domain would be consistent with the importance of cohesiveness in problem domains [7]. Domain rules serve as meta-rules that govern what subordinate categories of requirements are appropriate for the domain. In this way, domain rules are constraints. Examples of domain rules for the two domains considered in *Breakdowns* are listed in Table 16 and Table 17.

As noted above (see Internal Controls section), domain rules are stable over time, in contrast to business rules (see Business Rules and Volatile Variability). Domain rules are unlikely to change over the entire life of the system, whereas business rules may change frequently over the system's life. Given this dichotomy between domain rules and business rules, domain rules are unlikely to ever become business rules or vice versa. Domain rules represent critical knowledge that, by their enduring nature may become tacit (see the Procedure section in Chapter 3). Domain Rules Analysis makes domain rules explicit (see Procedure section). Their stability is a defining characteristic of domain rules, arising out of their basis in the underlying principles, theory, longstanding practices, or traditions of the domain. Unlike the concept of a kernel in OODA [e.g., 94], commonality (functions performed by multiple applications within a domain) is not a defining characteristic of domain rules. Domain rules may represent either commonality or stable variability in a domain (see Procedure, step 15). Table 11 relates domain rules and business rules to commonality, stable variability, and volatile variability (also see related Table 5).

Table 11: Commonality, Stable Variability, and Volatile Variability

Category	Commonality	Variability	
		Stable	Volatile
Business Rules			X
Domain Rules	X	X	

Domain rules are the invariant rules for a domain. That is, all applications (systems) for the domain must take account of the domain rules to determine which apply. Not all domain rules apply to every application of a domain, but each application must incorporate at least one of the domain rules in order to belong to the domain. The differences in domain rules applicable to systems in the domain reveal a partitioning of the domain through stable variability.

Domain rules that represent commonality are a primary source of reuse in the domain, especially with OOT, where commonality can be incorporated through inheritance (see discussion of Figure 17 and Procedure, step 15). Derived classes would reflect variability, such as between cycles in the AIS domain (see Procedure, step 15 and Figure

67). This type of variability would be stable (see Procedure, step 15). In OOT, stable variability (Domain Analysis Background section for other approaches to dealing with variability) would be captured through specialization (see Figure 67) during design as part of the domain-wide framework (see steps 14 and 15), contained in the Meta-Artifact. These specialized classes would be selected from the domain-wide framework during composition [93 and 95], and activated as services (see Procedure section, step 15) at runtime. Appropriate patterns would be used at all three levels [49 and 84].

Stable variability would be a secondary source of reuse. By factoring out volatile variability (see Business Rules and Volatile Variability), and specializing domain rules, stable variability would potentially be applicable to multiple applications within a domain with little further specialization. Stable variability would be a source for distinguishing one application in the domain from another, but it would not be a source of volatility. For example, the choice of a depreciation method for a category of capital assets would be a business rule, subject to relatively frequent change based on regulations and policy. However, the code implementing a depreciation method – stable variability – is highly unlikely to change.

Using Domain Rules Analysis to produce a Bifurcated Architecture systematically deals with both commonality and variability. These two extensions (see Table 6) of MAP intersect. While Domain Rules Analysis leads to a framework for commonality from which stable variability can be derived through specialization, the Bifurcated

Architecture goes farther by separating volatile variability from both commonality and stable variability.

Business Rules and Volatile Variability

Business rules reflect the volatile variability of the domain, in contrast to the stability (and commonality in some cases) of domain rules and stable variability (see p. 154). The concept of volatile variability provides a means of identifying business rules. Business rules are volatile because they change in response to such things as the current situation – e.g., environmental conditions, technology, knowledge, and attitudes. Unlike domain rules, a system's business rules are under the control of the organization that uses the system. They capture the organization's business philosophy and practices in terms "... that describe, constrain, and control the structure, operations, and strategy" of the organization [126, p. 5]. Business rules may be derived from external sources that the organization does not control, such as domain rules, regulations, and cultural considerations. Such business rules reflect the organization's interpretations of how to comply with external sources. The organization's interpretation may change both in terms of how to comply and of which external sources are relevant. To the extent that the external source is itself subject to change (e.g., frequently revised federal regulations), the volatility is increased.

Volatile variability supplements the customary definitions of business rules by placing business rules in the context of commonality and variability and aiding in their identification during all phases of the system's lifecycle (see Definition of Domain Rules

Analysis section and Procedure section, steps 5 and 10; see p. 152). That is, developers and stakeholders can apply the criterion of volatility – frequency of change – to identify functions to be externalized (see Procedure section, step 16). The volatility criterion should make it unnecessary to have detailed definitions and rules for identifying business rules. While ultimately a matter of judgment as to whether the variability was volatile, the likely frequency of change would provide an objective measure. The frequency might vary from very often (e.g., hourly or daily for online sales or time critical targeting) to a few times over the lifecycle of the system, but regardless of the exact frequency, when business rules do change, externalizing them avoids costly maintenance activities (see pp. 155-157).

Business rules may be viewed as a constraint on particular applications in a domain. They may also be viewed as the decision-making rules for the application, within the invariants of the domain (represented by the domain rules). Because business rules include the policies of the enterprise [99, 100, 125, and 248], they are the same rules often needed by decision support tools [211] to analyze the effectiveness of the enterprise's operations, especially for what-if scenarios based on changing policies. In this light, business rules represent institutional knowledge that may be tacit, embodying unstated and taken-for-granted assumptions underlying organizational practices. Making business rules explicit and accessible to all authorized stakeholders would help reverse this knowledge entropy (see Meta-Artifact section and Appendix B, Meta-Artifact and Time) to increase understanding of the system and future requirements for the system or new target systems in the domain.

Separating business rules from program code for other requirements would contribute to reuse (see Table 27), because business rules would, by design, capture the volatile requirements. The program code would be reserved for capturing requirements that seldom change (see Domain Rules section, stable variability).

A major decision to be made by the developers for the business rules would be whether to allocate the business rules to their own class hierarchy, linked by appropriate associations to the remainder of the framework (see Figure 91 for an example of such a separate hierarchy for business rules). Such separation may not be necessary or even appropriate in all cases, but it seems appropriate for the AIS domain, where business rules readily fall into traditional categories (see Appendix D). Figure 68 shows how polymorphism would be used based on an inheritance hierarchy for business rules.

Business rules are not likely to be documented in the primary sources for the domain (see Procedure, step 1) or even the supplemental sources (see Procedure, step 2) of a domain [99 and 248]. Because of the volatility of business rules across time, organizations, and missions, they would likely consume most of the requirements gathering effort for a bifurcated architecture, for reasons such as the following:

- Interviews with current users and domain experts would be needed to obtain the current business rules (this is generally the most time-consuming and costly step in auditing, for example, creating a significant burden on organizations)
- There would likely be a large number of business rules

- Written sources would be less readily accessible than those for domain rules or the supplemental sources in that they would exist in private and/or protected repositories, such as organizational policies and procedures manuals, or embedded in legacy computer applications and personal knowledge [99 and 100].

- There may be discrepancies between the business rule or the prescriptive instructions (see Internal Controls and Table 20 and Table 21) to carry out the rule may change

Business rules may apply to automated functions or people. If they applied to people, as would most ROEs in Table 21, then they might affect the interfaces with supporting automated systems. Such automated support may be indirect, for example, decision support aids for calling in direct fires. Rules that typically apply to people could also form the basis for automated functions such as robotic behavior in combat, target identification in missiles, or weapon-target pairing in manned aircraft, with related use in data correlation and fusion algorithms (see 151 and 282).

Definition of Domain Rules Analysis

The following definitions take account of how Domain Rules Analysis focuses on the problem space so that a comprehensive domain analysis is performed to at least satisfy the requirements of the domain rules, before considering the solution space. That is, analysis of the domain continues after the requirements for the initial system have been analyzed (see Figure 16 and Figure 17), if there are domain rules for which the requirements have not yet been analyzed, to produce domain-wide frameworks and patterns for adaptation and extensibility (see Appendix B, Adaptability and Extensibility).

> A domain is the set of requirements for one or more systems within boundaries established independently of the solution space, that is, the requirements would clearly be independent variables.

This supports the following definition:

> Domain Rules Analysis is the process of studying a domain to identify its domain rules, then the commonality and variability among systems that would meet at least the requirements of the domain rules.

This refinement of the traditional definition of domain analysis theory assumes that there is a set of stable rules for the domain – domain rules –, which can define the boundaries of the domain.

Because of the concentration on the entire domain in Domain Rules Analysis, any established information about the problem domain could supplement the domain rules in defining the domain, as well as determining how to support the domain rules. Chapter 3 describes a detailed Procedure for performing Domain Rules Analysis. As discussed in the Procedure section, supplemental information would also assist in discovering tacit domain rules. Two supplemental sources of domain information (see Procedure section) would be particularly useful in this regard:

- Traditional subdomains, functions, methods, processes, and procedures
- Established processing cycles, business processes, or patterns

The detailed discussion centered on Figure 16 and Figure 17 below describes how the dynamics of Domain Rules Analysis contrasts with current domain analysis in maintaining such a focus. Domain Rules Analysis is depicted graphically in Figure 17.

Figure 16 shows how analysis of requirements from the problem space is typically limited to a solution (yellow – IR triangle in Figure 16) for the initial domain customer's requirements (yellow – IR circle in Figure 16). A second solution (tan – FE square in Figure 16) for the next domain customer's requirements (tan – FE circle, around IR circle), often assumes all or some of the first solution is applicable without a thorough analysis of the second customer's requirements. Subsequent solutions tend to rely increasingly on existing solutions – less on a thorough-going requirements analysis – sometimes accompanied by efforts to generalize the solution if additional customers are anticipated, e.g., as an object-oriented framework (blue – SE grid). The blue SE grid is an instance of what Cohen described as evolving a framework (see p. 143) as systems are developed.

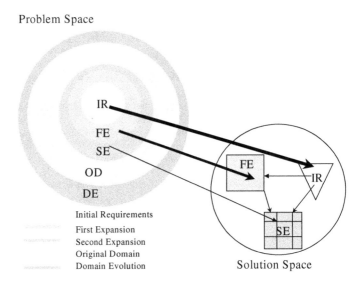

Figure 16: Typical Analysis of Requirements

This progression toward less emphasis on the problem space is shown graphically in Figure 16 by the narrowing lines from the problem space to the solutions in the solution space. Likewise, the arrows from the first solution to the second and from the first and second solutions to the third solution show the increasing reliance on solution-space analysis.

Domain Rules Analysis first identifies domain rules (see Procedure section, steps 1, 2, and 3). Domain Rules Analysis then keeps MAP focused on the problem space (domain) by using domain rules to identify use cases independently of the solution-space (see Procedure section, step 1.3), before considering solution-oriented sources of domain information (see Procedure, steps 2 and 9). Use cases are problem-oriented by definition (see Unified Modeling Language section), dealing with the requirements (explicit statements of needs) as the independent variables of the problem space, reinforcing the problem-oriented focus of Domain Rules Analysis (see p. 143).

While focusing on the problem space, Domain Rules Analysis provides a comprehensive analysis of the domain, using domain rules as a measure of completeness (see Procedure section, steps 1, 7, and 11), ensuring that the requirements of the domain rules, at a minimum, have been identified and analyzed. During this comprehensive domain analysis, Domain Rules Analysis also categorizes requirements that represent volatile variability as business rules (see Procedure section, steps 4). The Bifurcated Architecture (see p. 149), requires that volatile variability be identified throughout the application of the MAP (see Procedure section, steps 4, 5, 10, and 16).

Figure 17 depicts the Domain Rules Analysis of the original domain (green – OD circle) to create a framework for the entire domain (Domain Rules Framework in Figure 17). All of the particular solutions are derived as adaptations from the Domain Rules Framework, through specialization (by using object-oriented technology, the frameworks are various class hierarchies; see Appendix B, Framework). The arrows in Figure 17 show this progression from problem space, to framework, to solution space graphically.

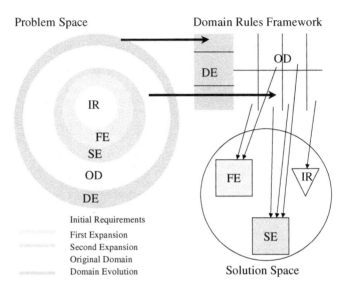

Figure 17: Domain Rules Analysis of Requirements

The blue – SE square representing the third solution is no longer shown as a grid, which indicated in Figure 16 that it was a framework. Individual solutions do not need to be generalized, because adaptation for reuse is achieved through the Domain Rules Framework.

The purpose of performing the Domain Rules Analysis would be to produce a framework for the domain that was sufficiently complete that new base classes would be needed only when the domain evolved (see p.148). The proportion of derived classes available initially would depend on the domain and circumstances, such as priority applications and funding (see Figure 17). In a bifurcated architecture, once the stable (over time) domain framework and patterns (steps 14 and 15) were developed for a domain, specific applications would be composed from components derived from the framework, guided by the patterns. Tailoring of applications could focus almost entirely on the business rules.

Over time and across applications, the classes in the framework could be reused (see Table 27) with few changes, for two reasons:

- The completeness of the domain-wide framework created using Domain Rules Analysis
- The stability of the classes

When new derived classes had to be specialized for adaptation (see stable variability in the Domain Rules section and Appendix B: Adaptability) or new base classes because of domain evolution (Appendix B: Extensibility), the new classes would themselves become highly reusable assets, for the same two reasons

In order to justify the effort required by Domain Rules Analysis, an organization would have to see a need to provide a domain-wide framework – "... an abstract design ... in the form of a set of classes [49]" – along with patterns to satisfy its internal plans, a perceived market, or some combination. The organization would then have to invest substantially in the development, probably at considerable risk, helping to explain why such a complete domain solution is uncommon.

Domain Rules Analysis can be especially useful in a network-oriented environment, where real-time, distributed, concurrent systems are essential, with the emphasis on the software components [22] of the system and how they collaborate (network is a group of nodes and the links interconnecting them [124]), rather than the platforms that hold the components or algorithms and data inside them. This emphasis, which parallels the definition of system architecture (see the section Architecture Centricity below), also highlights architecture centricity, providing another example of the interdependence of the elements of MAP (see Dynamic Interdependence of the Eight Elements of MAP section).

Because Domain Rules Analysis provides a process for performing a comprehensive examination of the domain, by applying it in close coordination with two of the capabilities of the enabling technologies – one for analysis and one for implementation – it can effectively address the needs of network-centric environments. The analytical capability is the collaboration diagram (see Object-Oriented Technology section below). The implementation capability is the integrated tool support for publish and subscribe interfaces, which preserves the integrity of the collaboration analysis (see the Unified Modeling Language and Integrated Modeling Tools sections below).

During development, Domain Rules Analysis can be used to systematically identify the interfaces required among the objects in their collaborations. Such a focus on the interfaces as well as the structural and behavioral elements of the domain would contribute significantly to the completeness of the Meta-Artifact. These interfaces could then be generated as publish-and-subscribe interfaces by the toolset when the target system is generated. Because collaborations show the connection pattern – links – among a set of objects – nodes – to accomplish a specific purpose [122], they parallel both the definition of a network [124] and a system architecture (see Architecture Centricity section below), collaborations provide another example of the interdependency of the elements (see Dynamic Interdependence of the Eight Elements of MAP section).

Impact of Domain Rules Analysis

Domain Rules Analysis, by providing a method for identifying volatile variability (see Chapter 3, Procedure section) for the Bifurcated Architecture, also contributes to the impact on developers of the paradigm shift in how they view the system with MAP (see Role of the Meta-Artifact in MAP section). Domain Rules Analysis further contributes to the Meta-Artifact and MAP by allowing a more complete and precise definition of a domain by identifying its domain rules (see Chapter 3, Procedure section). To the extent that patterns and frameworks of classes can be derived with Domain Rules Analysis to completely cover the domain (see Figure 16, Figure 17 and related discussion), the Meta-Artifact subsumes the solution space, where target solutions – including the running system – are just views of the Meta-Artifact (see p. 121). These views may initially be at the level of architectural frameworks and patterns [21, 22, 49, and 84], which may need to be adapted or extended (see Domain Rules Analysis and Appendix B, Adaptability and

Extensibility). In this sense, the Meta-Artifact represents the totality of the solution space.

Impact on control needs – In making the volatile variability of the system directly accessible to end users and other authorized stakeholders, the application of Domain Rules Analysis to produce a Bifurcated Architecture has a direct impact on controls (see *Impact on Control Needs* in the Bifurcated Architecture to Encompass Commonality and Variability section).

Bifurcated Architecture to Encompass Commonality and Variability

The Bifurcated Architecture separates the volatile variability – business rules – from stable variability and commonality (see Appendix B). Developers allocate the prescriptive rules implementing the business rules (see Appendix B, Controls) to an external repository (file or database; see Figure 18, Figure 27, and Figure 50; pp. 154, 153, and 280) where they can be maintained more efficiently and accurately than when they are embedded in code, because they are treated as data by the system. As external data, the business rules become more accessible for such purposes as analysis, audit, and training. In this way, the Bifurcated Architecture greatly simplifies adaptation in the current domain and extensibility in an evolving domain, because the features most likely to change – represented as business rules – can be easily modified, even in realtime.

The Bifurcated Architecture enhances the OODA concepts of commonality and variability in two ways:

- Differentiating stable (see Appendix B) variability from volatile variability

- Providing physical as well as logical separation (Figure 18, Figure 27, and Figure 50)

The separation provided by the Bifurcated Architecture is both logical and physical, yielding multiple benefits, including:

- Improved reliability by eliminating the need to change software code as is done when business rules or rules of engagement are not externalized

- Reduced maintenance costs by eliminating the need to change software code as is done when business rules are not externalized

- Increased visibility of the enterprises business rules for control, consistency, financial audits, and security audits

- Increased reusability and interoperability by removing volatile variability from the software code in components, so that the components represent only commonality or stable variability

- Use by decision support tools

Common definitions [see p. 59] of business rules do not include volatile variability as a defining characteristic.

Business rules (see Appendix B) have always been captured and analyzed with other requirements, whether implicitly or explicitly [99], but the approaches cited in the Survey and Review of the Literature of Prior Research section for handling business rules take one of four general approaches for the balance of development:

1. Extracting business rules from existing portions of applications for a narrow or particular function, purpose, or aspect and managing them externally to application code (e.g., credit worthiness, online sales support, autonomous behavior, or expert knowledge in decision support such as choosing a course of action during combat)

2. Using business rules as requirements for analysis or design, but not runtime

3. Developing entire rule-based applications with a rule-based programming language such as ProLog

4. Developing a business-rules-centric architecture

Of the four general approaches, number 3 would always be rule-based (involving artificial intelligence; see Appendix B, Rule-Based Process); 1 and 4 might be either rule-based or rule-constrained (not involving artificial intelligence; see Appendix B, Rule-Constrained Process). Neither rule-based nor rule-constrained would apply to general approach 2, which does not involve implementation.

The Bifurcated Architecture is rule-constrained, rather than rule-based, which distinguishes it from general approach 3. While an inference engine might manage the rules allocated to repositories, their management would not require the sophisticated capabilities of an inference engine, such as backtracking (see Inference Engine Considerations section). However, the separation and accessibility of the business rules would allow their use, in conjunction with the ontology provided by the Meta-Artifact (see Inference Engine Considerations section and Generation of Ontologies and Taxonomies section), for composing true expert systems for applications outside the normal production applications, e.g., for autonomous operation or decision support.

General approach 2 only applies to analysis and design, whereas the Bifurcated Architecture carries through implementation, operation, and maintenance. One result of this is that the identification of the business rules continues into the code. This is an example of how the concept of volatile variability aids in identifying business rules as the concept of domain rules aids in defining a domain: rather than focus on how to state businesses rules at the requirements level [99, 100], this extended definition allows business rules to be identified based on one of their defining characteristics – volatility.

Coupled with Domain Rules Analysis, Bifurcated Architecture seeks to systematically and completely analyze a problem domain in a way that identifies its volatile variability along with other outcomes of the analysis, regardless of what the basic function, purpose, or aspect of the resulting components may be, in contrast to general approach number 1.

Business-rules-centric approaches (number 4) focus the analysis on rules - their definition, role, interaction, and validity within the organization - then base an architecture on a central database of rules or particular aspects of the system that rely heavily on rules. In contrast, the Bifurcated Architecture is a result of MAP, an architecture-centric approach, so the bifurcation is orthogonal to the system architecture.

Architecture centricity emphasizes control, interoperability, reuse, and reliability (see p. 175; Table 27).

Apart from separating the volatile variability at the point of occurrence (see Figure 18, Figure 27, and Figure 50; p. 281), the architecture is unaffected. The architecture-centric MAP proceeds to identify an architecture that is most appropriate for the domain requirements - it is not driven by business rules, i.e., not business-rules-centric. The difference is fundamental and philosophical. MAP does not impose a particular architecture or specific design patterns. It starts at the beginning, with the problem space. MAP leads to architectural patterns (frameworks) and analysis patterns (collaborations) for the problem space. Design patterns would follow. A major concern of MAP is to avoid imposing solutions that have not been derived from the problem space. Once enough analysis work has been done to discern patterns for the problem space, developers would refer to sources of existing patterns to find matches. They would do the same for design. Figure 18 and Figure 50 show implementation views of the Bifurcated Architecture. Figure 27 shows a design view, with business rule constraints tethered to design classes (classed sufficiently detailed to generate code). As suggested by Figure 27, not all classes (or components) would have rules. In this sense, the Bifurcated Architecture extends OOT, by adding greater encapsulation for rules.

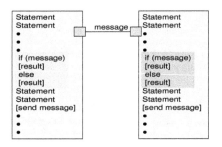

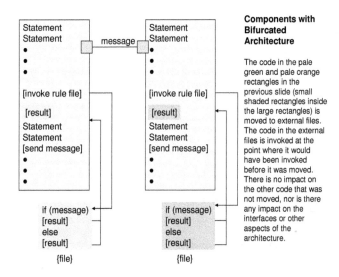

Figure 18: Bifurcated Component

(Related to Figure 18, see Steps 4 and 16 of Procedure for applying MAP in Chapter 3, separate volatile variability in all phases of development.)

Object-Oriented Technology facilitates separating business rules from the rest of an application in a decentralized manner for the Bifurcated Architecture, because the prescriptive rules implementing the business rules can be allocated to repositories (see Figure 18, Figure 27, and Figure 50; Table 20and Table 21; pp. 137, 157, 259, and 281) corresponding to individual components, if appropriate for the architecture. The concept of the external business rule repository does not depend on a centralized database. The defining characteristics of the Bifurcated Architecture are that prescriptive rules implementing business rules are:

- Separated from program code, hardware logic, or something in between such as Field Programmable Gate Arrays
- Accessible to all authorized stakeholders
- Changeable by all authorized stakeholders

Allocating the business rules to a database apart from the code provides two major benefits. First, the aspects of the system that are exposed to frequent change can be isolated for operations and maintenance. Once isolated from the software code, business rules can be maintained through mechanisms appropriate to their volatility, accessible to end users, rather than the traditional methods for maintaining software code. Using these traditional methods the cost of writing code to enhance or modify a legacy system can easily be three times as expensive, with three times as many defects, as writing code for a new system, and the probability of introducing a new error when fixing a defect can be as high as 60% for complex code [136, pp. 330-333]. By allowing rules to be changed as data in an external database, the Bifurcated Architecture contributes to reliability by reducing the likelihood of introducing defects through the error-prone process of changing code, by as much as a factor of three, plus a significant reduction in errors introduced while fixing those defects (see p. 282 for errors that might be introduced by end users).

Second, all of the decision rules used by the system are accessible to any authorized stakeholder without specialized engineering support. By making the business rules accessible in this way instead of leaving them embedded in software or hard-to-access documents, stakeholders would be able to readily assess the organization's operating philosophy and policies. The business rules database would be a key component of the information duo described in [74], which would give stakeholders a comprehensive picture of the organization in electronic form.

Bifurcated Architecture and OOT intersect by using domain rules and business rules to amplify the generalization and specialization of object-oriented inheritance and to provide criteria for commonality and variability in domain analysis. That is, the commonality identified through Domain Rules Analysis (see Definition of Domain Rules Analysis section and Procedure section) can be captured in the parent classes, with specialization applied for stable variability to deal with differences among target systems (see Domain Rules). Factoring out the volatile variability of business rules increases the commonality of parent classes and the stability of the specialized classes by removing aspects of variability that would otherwise be embedded in the software code (see Figure 18, Table 27).

In concrete terms, the business rules would be maintained in a runtime rules file that would be processed by an inference engine (see p. 259 and Appendix C for implementation discussion and details). These rules would be accessible to authorized operators for maintenance or information purposes as well as for operation of the system. Operators would maintain the rules through a Human Machine Interface (HMI), rather than requiring the lengthy process for code changes requiring an engineer in the traditional approach to changing business rules, e.g., requirements, analysis, design, implementation, testing, transition, and related approvals (see pp. 155 and 157). COTS software is available today to accomplish this rules maintenance.

Over 60% of the lifecycle costs of software intensive systems are due to enhancements and defect repairs [136, p. 319]. As the most volatile component of software, business rules would account for a major share of the enhancements and repairs (see pp. 155). In a military context, for example, these include rules of engagement, mission plans, and resource management. But this only captures actual costs, not the opportunity costs that are incurred because rules are not changed when they should be because of delays or expense.

The second major benefit of the business rule database – accessibility of all of the decision rules used by the system to any authorized stakeholder without specialized engineering support – supports the following uses:

- Audits (e.g., security or financial)
- Analysis for changes to the rules
- Recursive use to check new rules before they are accepted into the database (e.g., to avoid inconsistencies with other rules)
- Use by decision support tools

Impact on control needs – Bifurcated Architecture, through this increased reliability described above in this section, addresses the problem of preserving the integrity of controls while reducing maintenance costs over the lifecycle of the system (see pp. 155-157). In particular, many of the control needs of a system are captured in business rules (see Internal Controls section and pp. 220 and 281). By their nature, controls must test facts to determine if prescribed results have been met. That is, controls must enforce the constraints of the system. The controls are the prescriptive implementation of the business rules (see p. 61 for the particular type of controls implemented in the prototype).

Making the volatile variability of the system directly accessible to end users and other authorized stakeholders also plays a major role in making knowledge explicit that would otherwise be embedded in the system as software code and become tacit (see Domain Rules Analysis section above). Making knowledge of the system and the organization it supports explicit so that it does not become part of the tacit background also avoids breakdowns. Failure to articulate such tacit knowledge as artifacts is a cause of breakdowns in meeting control needs (see the Background of the General Problem section above).

Impact of Bifurcated Architecture

The Bifurcated Architecture directly affects how practitioners view their field of endeavor. The Bifurcated Architecture departs from other concepts of system implementation in how it deals with commonality and stable variability on the one hand, and volatile variability on the other. The collecting of business rules that reflect the volatile aspects of a system (the volatile variability identified by Domain Rules Analysis) into an external repository rather than embedding them in software code requires radical shifts in thinking that may lead to resistance based on lack of understanding or agreement, as discussed in Contrasting the Meta-Artifact with Artifacts of Other Processes. This will be a significant barrier to some. Different stakeholders may take opposing views on the costs and benefits. Business analysts may see great benefits with little cost. Programmers may see few benefits because the traditional approach of dispersing and deeply embedding business rules in software code is compatible with their mental models for programming [254], and they may perceive their role as diminished by the application of a bifurcated architecture. These are specific examples of the more general cultural impact discussed in the Cultural Impact section.

Object-Oriented Technology

The defining features of OOT enable or amplify the properties of the Meta-Artifact (see p. 74). Table 14 summarizes many frequently noted benefits of OOT as reflected in the

Meta-Artifact. For a broader perspective, Alexander Levis in his course book for C4ISR Architecture Framework Implementation, observes, "The world is moving toward object orientation" for software. This has overarching implications for supporting systems, including:

- Avoiding obsolescence
- Using the growing base of developers who can support OOT
- Addressing a growing audience of stakeholders who can understand OOT, with UML as the standard representation

The Meta-Artifact leverages the emphasis placed on standard interfaces by encapsulation – a defining property of OOT – augmented by publish-and-subscribe services, to assure the interoperability of components. Because the components are object-oriented, the functions are encapsulated in objects within the components. This encapsulation limits access to functions to services published by the standard interfaces. Focusing on the services and using specialization through inheritance facilitates cohesiveness within objects and components. These same qualities – standard interfaces, cohesiveness, and access to published services – directly enable distributed, concurrent processing. They also strongly support the use of the Meta-Artifact for enhancement of implemented systems and implementation of new target systems through composability, guided by patterns defined in collaborations of encapsulated objects.

Inheritance – another defining characteristic of OOT – also promotes reuse of the Meta-Artifact by capturing commonality and variability in base and derived classes (see Domain Rules Analysis section).

Polymorphism – the third defining property of OOT – in combination with encapsulation and inheritance, contributes to extensibility [61], reliability, and security [146].

A collaboration is the central analytical concept in Object-Oriented Technology (OOT) [21, 22, and 200] – as noted above, the collaboration diagram appears in all five UML views (Figure 19). The Unified Modeling Language (UML) provides a standard representation for such collaborations. Integrated modeling tools automate the production and manipulation of collaborations to generate software code and hardware designs to implement them. The integrated support for publish and subscribe interfaces allows the software components generated by this code to interact across a network just as the collaboration-based analysis prescribes. The close conceptual relationship between a collaboration and a network (see pp. 161 and 147), allows the analysis from one to the other to proceed with an intuitive mental transformation [254]. Because of this conceptual and practical synergy, object-oriented concepts, UML, and related COTS tools provide a natural means of developing network-centric systems.

Because classes and objects in a collaboration may represent any aspect of a system – processes, data, concepts, things, or some initially ill-defined combination – collaborations among them can be used to capture the semantics of the system throughout development. Depending on the level of granularity:

- Processes may consist of few or many responsibilities [22], ultimately through analysis, design, and implementation to become computer instructions
- Data may be uninterpreted patterns, information based on interpretation, or knowledge based on understanding of information
- Concepts, such as a customer account or commander's intent
- Things may be large composites such as a ship or an organizational unit, or small components such sensors or computer processors

Eventually, things may serve as platforms for objects (i.e., other things, processes, etc.). There may be collaborations among such platforms and among the objects on the platforms. However complex this may become, collaboration-based analysis provides a simple building-block approach that is intuitively understandable to all of the stakeholders, contributing to the Meta-Artifact's quality of promoting understanding of the system (see pp. 187 and 221). Because these collaborations are actively linked to the running system, the semantic content of the collaborations gives the stakeholders a correct and current understanding of the system.

Impact on control needs – Together, the three defining characteristics of OOT contribute to a system's control needs: encapsulation protects the code implementing controls; polymorphism, in concert with inheritance, encourages the use of controls declared in the base classes of the framework, by derived classes throughout the framework. Inheritance, through use of pure virtual functions (see pp. 283 for use in prototype) – to create an abstract class – and derived classes, can assure that certain functions are always carried out by derived classes as prescribed by the base class. This in turn is enhanced by polymorphism, which allows different derived classes (forms) to respond differently to the same base-class function call by altering the base-class function according to their

derived-class definitions, e.g., those enforcing security policies [61]. Polymorphism also

reduces the number of function names, which contributes significantly to understanding

and error reduction.

Unified Modeling Language

UML has become the standard representation for object-orientated software [4, 75, and

181]. UML provides the common intuitive visual representation for artifacts from

requirements elicitation (Figure 21) through analysis (Figure 28) and design (Figure 22),

that is, from use cases (descriptions of what the system is to do in terms of observable

results [22]) through collaboration and sequence diagrams. This consistent representation

of UML modeling even carries through to the structure of code in object-oriented

programming languages, such as C++. Just as a UML class icon has compartments for

class name, attributes, and operations, source code for C++ is organized into sections

identified with a class name, followed by attribute and operation declarations.

Using the development tools that implement UML, developers can produce a model that

provides multiple views of the system, one of the properties of the Meta-Artifact (see p.

74), as described in the following excerpt from IEEE 1471 "Recommended Practice for

Architectural Description of Software-Intensive Systems:"

> In the conceptual framework of this recommended practice, an
> architectural description is organized into one or more constituents called
> (architectural) *views*. Each view addresses one or more of the concerns of
> the system stakeholders. The term view is used to refer to the expression
> of a system's architecture with respect to a particular viewpoint.

NOTE—This recommended practice does not use terms such as *functional architecture,* *physical architecture,* and *technical architecture,* as are frequently used informally. In the conceptual framework of the recommended practice, the approximate equivalents of these informal terms would be *functional view, physical view,* and *technical view,* respectively.

Figure 19 shows graphically the concept of multiple views of a system architecture in UML. Within the categories of views depicted in Figure 19, stakeholders can see varying levels of detail. From these views, stakeholders, usually developers, can create multiple models [21], e.g.:

- Domain model that establishes the context of the system
- Business model that establishes an abstraction of the organization
- Use case model that establishes the system's functional and non-functional requirements
- Analysis model that establishes an idea or conceptual design

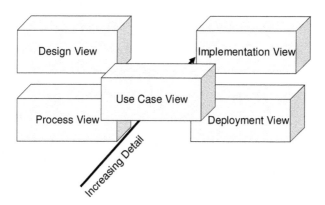

Figure 19: Multiple Views

Domain models and business models are just different views of the system architecture, as captured in the Meta-Artifact. In turn, a model has multiple views, where each is a projection into the model [21]. That is, each of the views in Figure 19 may consist of projections into multiple models. This means that requirements gathering and analysis can start at the highest level, examining the entire enterprise (the business model) and its interaction with its environment when appropriate. Whether at the level of the enterprise or a subdivision of it, UML supports end-to-end modeling for the organization and its environment, using the same notational representation at all levels. This end-to-end capability supports the gathering of comprehensive requirements (see Domain Rules Analysis), i.e., by starting at the right level and preserving the requirements through the active semantic chain created by the actively linked artifacts of the Meta-Artifact. The common representation is essential to the multiple-view property of the Meta-Artifact. These views are actively linked to the visual model from which the code for them is automatically generated, providing efficient and effective testing through Visual Verification and Validation (see Animated, rather than static, views of development artifacts section). The common representation is also essential for the Meta-Artifact to avoid the gaps discussed in Contrasting the Meta-Artifact with Artifacts of Other Processes.

The five architectural views shown in (Figure 19) become successively more concrete in the following sequence:

1. User

2. Design
3. Process
4. Implementation
5. Deployment

Because UML is oriented toward iterative development, many of the same diagrams are used in multiple views, e.g., collaboration diagrams are used in all five views. The collaboration diagram has a central role in UML-based development (see p. 161). With iterative, incremental methods, the distinctions among views are not so much time-dependent as content- and level-of-detail dependent – contributing to the Meta-Artifact's being the totality of the system (see Impact of Domain Rules Analysis section) at each stage of the system's lifecycle. That is, views created with the disconnected artifacts described in Contrasting the Meta-Artifact with Artifacts of Other Processes, were time dependent in that the design view was not available early in the development process, say while requirements were still being gathered. The five views may exist at differing degrees of detail through all phases of the development life cycle – they are all just different views of the Meta-Artifact, to which detail is added in successive iterations (see Iterative, Incremental Methods section). As noted above, IEEE 1471 seeks to improve the overall development process by standardizing this concept of a system architecture – captured in the Meta-Artifact in MAP – with multiple views.

Because of the nature of object-oriented methods and programming languages, code modules, as previously noted (see p. 163), have the same look and feel as related elements in the visual UML model. UML provides the active semantic chain with a consistent representation from visual element through automatically generated code, yielding important advantages for both development and maintenance, in terms of reduced cost, faster response to changes, increased reliability, and portability. Another important advantage in terms of establishing a common understanding of the semantics of the Meta-Artifact for diverse stakeholders is that understanding UML diagrams typically does not require special expertise, because they are labeled with familiar terms, using a small number of icons that quickly become familiar to those who review the diagrams, providing a common means of communicating among the stakeholders. (An overview of UML visual modeling elements provided by numerous articles and publications is helpful, but in practice a very brief explanation for stakeholders, for say requirements elicitation, has proved sufficient).

The diagrams are intuitive, because of the nature of object-orientation, where what the stakeholders see corresponds to objects familiar to them [254]. That is, an element of an objective system, say a weapon, is simply represented in the diagrams as that weapon. An existing system – e.g., COTS, GOTS, or legacy – is represented in the same explicit notation. If that weapon or system communicates with a sensor, the sensor is represented as a sensor. A line represents the communication between them with words on the line describing the communication. For example, the Base Active Class in Figure 20 could represent a weapon type, with Derived Active Classes 1 and 2 representing particular versions of that weapon. The names on the icons would be changed to identify the weapons. Graphics resembling the weapons can also be substituted for the standard

UML icons. However, a disadvantage of such substitution is the loss of broader understanding based on the standard icons.

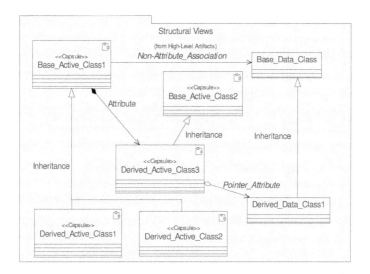

Figure 20: Structural Link in Semantic Chain

Integrated Modeling Tools

UML is a commonly used modeling language that many vendors support with a wide range of tools. IBM, for example, provides a complete set of integrated modeling tools that supports MAP, centered on its UML-based modeling and code generation tool, IBM Rational Rose Technical Developer (RRT). The other tools that integrate with RRT cover all aspects of the development cycle, including requirements management, configuration management, testing, performance optimization, and documentation.

These tools are tightly integrated, eliminating the need for duplicate efforts, so that the necessary artifacts for development – including documentation – are automatically produced as byproducts (see Special-purpose views and Active semantic chain sections). The configuration management tool, for example, leverages the inherent support of UML-based RRT (e.g., packages of artifacts that can be worked on independently) for geographically and temporally dispersed collaborative development.

In fact, the core tool, RRT, performs the basic functions of many of the integrated modeling tools, offering substantial flexibility in determining which tools are needed for a project, based on, for example, its size. The natural capabilities that UML and OOT provide for collaborative development such as packages and self-contained, encapsulated objects, is amplified through RRT's integrated publish-and-subscribe services. Iterative, incremental methods, described next, take full advantage of these capabilities and services.

A major advantage of a tool suite like that provided by IBM is that code is generated from the visual model – for a large selection of target platforms. The tool is the central code and architecture manager, a key factor for enabling:

- Adaptability – change within current domain (see Appendix B)
- Extensibility – for domain evolution (see Appendix B)
- Scalability – greater volume within current capabilities and adaptation (see Appendix B)

That is, the tool allows the insertion of external code from any source into architecturally

correct locations, then manages it along with automatically generated code, reducing

costs and preserving architectural integrity.

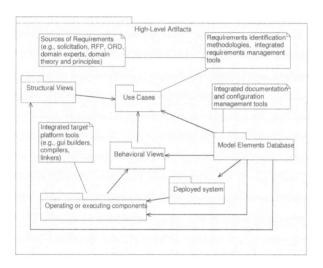

Figure 21: Beginning of Semantic Chain of Artifacts

Figure 20, Figure 21, and Figure 22 illustrate the role of integrated modeling tools in

generating the active semantic chain of electronically linked artifacts. Figure 21 shows

how this semantic chain begins, using the familiar shape of a folder to represent a UML

package, because UML uses packages to collect whatever elements, including other

packages, the developer chooses to organize the development project. The project

organization itself has multiple views: e.g., the project manager may have a view oriented

toward subsystem packages, whereas a developer of a subsystem may have views oriented toward packages for use-case realizations (references to use-case realizations or realization of use cases are based on the descriptions of such realizations in [22]), class structures, and behaviors. Arrows connecting packages show the relationships among packages, similar to cross referencing among physical folders for hardcopy documents. Figure 21 also shows the flexibility of the UML notation, with the text boxes attached to the packages. These three figures also demonstrate one way the semantics are linked from high-level to lower-level details, by automatically propagating the names of the packages (Figure 21), classes (Figure 20), and instances of classes (Figure 22), for example.

Ready access to the high-level semantics greatly assists developers in determining in subsequent iterations the detail that must be added and where, using the tools to capture the details, then managing them through their linkage to the high-level semantics to ensure they work together as intended.

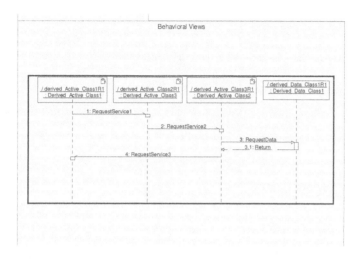

Figure 22: Behavioral Link in Semantic Chain

All of the diagrams are copies of the diagrams created by the tools from model elements in a database (Figure 11, Database). The current state of the individual elements and the relationships among them are available for viewing and development activities, by applying the appropriate integrated modeling tools to the database. Any detail added to the elements would be automatically propagated into any existing or future diagrams containing that element – maintaining global consistency among the elements – greatly simplifying iterative development, management of collaborative development, and stakeholder participation.

Comparing Figure 19 with Figure 21 reveals the flexibility of UML in providing multiple views. Figure 19 shows a generic developer's set of views, which support other stakeholder-specific views as needed, such as those in Figure 20, Figure 21, and Figure 22. There is considerable overlap in the diagrams used in the Logical and Process Views of Figure 19, varying largely in level of detail and diagram content (the elements such as class icons and relationship lines that appear on a type of diagram, such as a class diagram). This variation in depth of detail and content on a relatively small set of diagrams, as well as the organizational flexibility offered by the UML package concept, allows stakeholders to construct views that serve their purposes on any project, while limiting what they need to learn to understand the views.

RRT also supports extensive structured and free form annotations behind each diagram (such as the notation boxes in Figure 21). In the model, detailed textual information may be entered for the icons and connecting lines. These details are readily accessible by simply double clicking on the item of interest. The elements of textual information reside in the same database as the visual elements of the model so that they are accessible to integrated modeling tools to extract for documentation (Figure 11). Table 12 and Table 13 show samples from an Integrated Data Dictionary (CAF/DoDAF product AV-2) prepared in this way.

Table 12: Example of textual data extracted from Database of Model Elements

Node: Operational and Mission Planning – Node Data		
Data Name	**Data Type (Media)**	**Description**
ROEid	char	Table ID for rules of engagement
Target ID	char	Target identifier
TrgtNumberUnits	long	Number of targets
TrgtType	char	Target characteristics
TrgtDesignation	char	Target priority
TrgtLocation	char	Longitude and Latitude
WeaponID	char	Weapon selected for target

Table 13: Example of textual detail extracted from Database of Model Elements

Node: Operational and Mission Planning – Node Activities	
Activity Name	**Links for this Activity**
ReceiveROE	From BG/JTF-CDR
RequestTargetInfo	From BG/JTF-CDR From JTF-JIC From Theater Sensor
ReceiveTargetInfo	From National Sensors From JTF-CDR
SendMissionPln	To D & A

Architecture Centricity

Architecture centricity requires a process that preserves the integrity of the architecture and applies the architecture, as represented by the architectural baseline, for the entire lifecycle of the system. That is, architecture centricity is adherence to a particular architecture, once it is captured in the baseline; changes to the baseline should be minimal [22]. Architecture centricity strongly supports Domain Rules Analysis by providing a stable baseline as the analysis proceeds over many iterations and increments.

Architecture provides a unifying vision and guiding principles for a system [19 and 22]. The unifying vision and guiding principles provide an overarching concept that should be captured in the architectural baseline, to lead all stakeholders to a common goal. Architecture focuses on the significant structural elements of a system, their relationships, externally visible properties, and interactions among them. The need for a strongly controlled architecture can be viewed as a major cross-functional requirement. By realizing architecturally significant use cases [22 and 149] in the early iterations (see Iterative, Incremental Methods section) of development, stakeholders can agree on a stable architecture to provide an architectural baseline that will avoid rework later, while fully accommodating adaptation and domain evolution (see Domain Rules Analysis and Appendix B, Adaptability and Extensibility). Guiding principles, such as architecture centricity, are intended to govern the significant structural elements and the collaborations among them during the lifecycle of the system. For MAP, the guiding principles would include consistent application of the extensions and enablers (see Table 6).

The above discussion of architecture reflects both maturity over time and the widespread

acceptance of OOT. As the SEI notes on its Software Architecture Web site, "While

there is no standard definition [of architecture], there is also no shortage of them." The

following definition provided in [11] is typical:

> The software architecture of a program or computing system is the
> structure or structures of the system, which comprise software
> components, the externally visible properties of those components, and the
> relationships among them.

What are components, external properties, and relationships? In fact, relationships go

beyond structure and introduce behavior (interactions). The point is to allow flexibility in

these specifics, while taking a high-level view of a system, or as Garlan and Shaw

observe [369], software architecture is a level of design concerned with issues

> ...beyond the algorithms and data structures of the computation; designing and
> specifying the overall system structure emerges as a new kind of problem.
> Structural issues include gross organization and *global control structure* [italics
> added]; protocols for communication, synchronization, and data access;
> assignment of functionality to design elements; physical distribution; composition
> of design elements; scaling and performance; and selection among design
> alternatives.

Contemporaneously with the maturing concept of system architecture, the enabling

technologies – one of which is architecture centricity – were beginning to mature. The

emphasis of architecture centricity on the global attributes of a system, along with the

maturing of the other enabling technologies (see Table 6, Enablers), is crucial to creating

frameworks and patterns (see Appendix B) and composability (assembling components

into target systems [93 and 95]) of the Meta-Artifact. The current maturity of the

enabling technologies adds precision to the definitions of system architecture and provides the means to actualize them. The *UML Users Guide* offers this more detailed definition of architecture for software intensive systems:

> The set of significant decisions about the organization of a software system, the selection of the structural elements and their interfaces by which the system is composed, together with their behavior as specified in the collaborations among those elements, the composition of these structural and behavioral elements into progressively larger subsystems, and the architectural style that guides this organization of these elements and their interfaces, their collaborations, and their composition. Software architecture is not only concerned with the structure and behavior, but also with usage, functionality, performance, resilience, reuse, comprehensibility, economic and technology constrains and trade-offs, and aesthetic concerns.

The formality of UML, implemented with automated development tools such as RRT, gives precise, unambiguous meaning to each of the elements in this definition. The full realization of this definition by the architecture embodied in the Meta-Artifact, is seen through the multiple views of the Meta-Artifact that are supported by UML, but which require the integrated modeling tools for practical application (see Figure 6). By encouraging the concept of a single architecture for a system, with multiple views (such as those supported by UML) instead of multiple architectures, IEEE standard 1471 (see p. 163) also represents the degree to which the whole concept of system architecture has matured over the past decade, a relatively short time, as is typical of change in the information technology arena.

Impact on control needs – Garlan and Shaw's observation is especially helpful because it emphasizes the motivation for needing a new concept, more than trying to define the concept. It reflected the growing emphasis on reuse, reliability, and interoperability.

These three non-functional qualities of a system naturally moved the focus away from functional analysis with its concomitant emphasis on algorithms and data structures – typically represented as functional decomposition, logic flow, entity relationship, data flow [62], and control flow diagrams (examples of standalone artifacts with separate and distinct representations – see Contrasting the Meta-Artifact with Artifacts of Other Processes section). Support for these three non-functional qualities shifted the emphasis to the global attributes of a system, one of which is the global control structure. That is, one of the defining characteristics of architecture centricity contributes to a system's control needs.

Iterative, Incremental Methods

Using iterative, incremental methods provides strong support for accommodating domain adaptation and evolution (see Domain Rules Analysis and Appendix B, Adaptability and Extensibility) with MAP. MAP builds on iterative, incremental methods as described in [22], where development is divided into four phases (Inception, Elaboration, Construction, and Transition) and five core workflows (development activities – Requirements, Analysis, Design, Implementation, and Test). An iteration includes the five core workflows and a phase includes one or more iterations. In early iterations, developers realize architecturally significant use cases [22] to generate code (the increment for each iteration) and establish an architectural baseline [see p. 175]. Because of the automatic code generation, the Meta-Artifact provides stakeholders early views of the objective system, including hard-to-assess features such as the human machine interface (HMI) that provides early views of the objective system. Such early feedback serves to reduce risks by identifying defects early, when they are easily corrected, often in the next iteration. The feedback also reduces risk by providing concrete input to domain experts and other stakeholders as they revisit requirements, exposing missed requirements. Through successive iterations and increments, they add details to the model – from subsequent analysis and design, combined with realizing the remaining use cases.

Ultimately, all necessary detail for the objective system is added, through the basic steps of development, e.g., capturing requirements, analyzing them, and designing solutions. However, rather than gathering all requirements, completing the analysis, then designing the entire system, the basic steps are performed for each iteration, over short but complete cycles, in an integrated rather than a stovepipe-manner, to avoid the gaps between these steps noted in the section Contrasting the Meta-Artifact with Artifacts of Other Processes. This works synergistically within MAP to shift the development process from one consisting of a single sequence of disconnected steps to multiple sequences (iterations) of actively linked steps, continually filling in the details of the Meta-Artifact to expand the solution space without closing off later analysis of the problem space. If the domain adapts or evolves during development, the related requirements can be incorporated in later iterations, rather than reopening previous steps (e.g., completed requirements), because each iteration includes all of the steps of development.

Iterative, incremental methods intersect with Domain Rules Analysis by allowing stakeholders to view the evolutionary prototypes (see Kinds of Prototypes section) from their respective viewpoints, then adding anything that is missing to the next iteration, ensuring that all requirements within the boundaries of the domain – the problem space – are considered. Iterative, incremental methods also intersect here with both UML and the integrated development tools (e.g., RRT and its tool suite), which provide representations for multiple views and automation for the prototypes. The automation provided by the tools in conveniently realizing multiple views from the common database of system artifacts (see Figure 6) – for diverse stakeholders and levels of detail – is crucial. Equally crucial is the ability of the automated tools to maintain the active semantic chain.

Impact of MAP

The Meta-Artifact Process departs substantially from previous software development processes. As noted, it changes the way developers and other stakeholders think about a system (see Role of the Meta-Artifact in MAP section). This departure is based on the following:

- The contrast between the Meta-Artifact and the collection of artifacts produced in practice by other processes, as described in Contrasting the Meta-Artifact with Artifacts of Other Processes (shown graphically in Figure 12, Figure 6, Figure 7, and Figure 8)
- Role of the Meta-Artifact in MAP
- Bifurcated Architecture
- Domain Rules Analysis

Other aspects of MAP may contribute to this departure in that an understanding of OOT, integrated modeling tools, and domain analysis is also needed. In addition to the impact of MAP discussed in the sections Contrasting the Meta-Artifact with Artifacts of Other Processes and Role of the Meta-Artifact in MAP, the impact of the other two extensions is discussed in the next two sections.

Impact on control needs – The three extensions of MAP (see Table 6 and Figure 6) have a direct impact on control needs. The thorough analysis of the problem space encouraged by Domain Rules Analysis includes control needs. Allocating all domain rules to use cases, either as functional or non-functional requirements (see Procedure section in Chapter 3), provides a checklist for any related controls as the use cases are realized (see p. 220). Because of the nature of domain rules (see Domain Rules Analysis section below) in defining the domain, such related controls are likely to be architecturally significant [22] and have widespread use in target systems. Once the domain has been adequately analyzed, to provide sufficient completeness to identify a comprehensive control structure, the problem becomes how to maintain such a structure over the entire lifecycle of the system without loss of integrity, especially with respect to the original systematic analysis that led to the strongly controlled architecture baseline (see Solution Approach for the Constrained Problem and Architecture Centricity sections). Bifurcated Architecture addresses the problem of preserving the integrity of the controls while reducing maintenance costs over the lifecycle of the system (see pp. 155, 157, and 157). The third extension, the Meta-Artifact, facilitates the first two, as noted above in the section Dynamic Interdependence of the Eight Elements of MAP) emphasizing the interaction among all three and the enablers.

Cultural Impact

The different perspectives and roles for developers (see the Role of the Meta-Artifact in MAP section above) may create the kind of culture shock Kuhn describes [152]. One reaction, analogous to those noted by Kuhn as part of the culture shock created by paradigm shifts, is the discomfort or dismissiveness toward thinking about the Meta-Artifact, including the operating or executing components, as representing the actual system. Engineers who have spent their entire careers describing real systems in terms of self-contained and static artifacts have difficulty accepting the reality of a system without them (von Halle cites a similar cultural phenomenon in [248]). The conclusion of such engineers may be that if there is no System Specification Document, or a bound set of engineering drawings adhering to a prescribed format, there is no system. Perhaps the nature of the Meta-Artifact, especially its capacity to dynamically generate the objective system, makes it appear too ephemeral, lacking the static concreteness of a "real" system.

The idea of delivering a running system – the increment produced in an iteration – before analyzing or even gathering all the requirements is uncomfortable, incomprehensible, or unacceptable to many traditional developers. The idea of having a single architecture that subsumes the views of multiple engineering disciplines or business functions encounters similar objections. Likewise, the idea of generating an objective system from a visual model, one that is electronically linked to high-level requirements, strikes many as wishful thinking. These difficulties with new ways of dealing with a system – e.g., an overarching architecture, automatic code generation from visual models, early delivery of running systems, and an active semantic chain – are amplified when applied to an entire domain, as MAP does.

Another of Kuhn's observations, that what were problems or anomalous facts in the old paradigm become assumptions or logical inevitabilities in the new, also applies to MAP. Relatedly, there are different questions or different roles for old questions. An example is the central role of system architecture in MAP, as discussed in Object-Oriented Technology.

Table 14 contains examples of goals of systems development that are assumed or logically inevitable outcomes of MAP, but which are recognized as problems or hard to explain facts in the traditional development paradigm, when it appears that certain goals are never met. Many of these goals are elaborated by Pressman [200 and 200], as well as by Coad and Yourdon [47 and 48]. While the second column of Table 14 lists the primary elements of MAP that assume or logically lead to the goal, the elements of MAP are so interdependent that some contribution could be traced to each of them and this interdependence amplifies the impact of the individual elements in achieving the goals.

Table 14: Goals of Systems Development

Systems Development Goal	Primary Elements of MAP that Achieve the Goal
Development artifacts should be intuitive and mimic the problem domain to improve understanding of the system among the	OOT, represented in UML, captures real-world objects in ways that are intuitive (see p. 167)

Systems Development Goal	Primary Elements of MAP that Achieve the Goal
stakeholders, by providing a direct mapping from the problem domain to the representation of the system [47 and 48]	
Development artifacts should have a consistent representation [47 and 48]	UML provides a consistent representation for artifacts in all phases of development
Requirements should be traceable from each artifact [200]	The Meta-Artifact, actively links all development artifacts, using consistent representation, with integrated modeling tools
A system should have a single architecture with multiple views (IEEE 1471)	The Meta-Artifact, actively links all development artifacts, using consistent representation, with integrated modeling tools
The system development should be a process of refinement [200]	Refinement is a defining characteristic of iterative, incremental development, which is amplified by the properties of the Meta-Artifact
The systems development process should support partitioning [200]	UML packages and class structures systematically support partitioning is
The systems development process should produce modular systems [200]	Modularity is a fundamental characteristic of OOT
The systems development process should produce composable components [200]	Frameworks, patterns, integrated modeling tools, and target platform generation enable composability
Modules and components should have functional independence [200]	Functional independence is inherent in OOT (encapsulation) and publish and subscribe interfaces allow component interaction only through well defined interfaces
The systems development process should be architecture-centric[200, 21,22]	Architecture centricity is a defining property of UML
The systems development process should systematically account for non-functional requirements and properties [23, 22, 36, 149]	Use cases capture non-functional properties, which are allocated to model elements during use-case realization. Evolutionary prototypes produced through iterative, incremental development allow assessment of the non-functional properties.
Control hierarchy should express both visibility and connectivity [200]	Intuitive graphical representation of active classes (those classes controlling the processing flow of the system), their state machines, message flows, and interfaces establish a control hierarchy. Class icons, including interface classes, and class diagrams directly show visibility. Class, collaboration, and sequence diagrams directly show connectivity.
Internal details of a component should be hidden [200]	Information hiding is a defining property of encapsulation
The systems development process should produce accurate, complete, and current documentation	Current documentation is one of the qualities of the Meta-Artifact
The systems development process should reduce the risk of failure in meeting schedules and other requirements [200]	Iterative, incremental development, integrated modeling tools, and code generation contribute to reduced time to initial delivery and risk of failure.
Modules and components should be cohesive [200]	Encapsulation in classes derived through Domain Rules Analysis implement cohesiveness throughout the domain framework.
Coupling should be minimized to reduce the	Visual analysis of collaborations and interfaces,

Systems Development Goal	Primary Elements of MAP that Achieve the Goal
risk of propagating defects during development and maintenance [200]	implemented by encapsulation and published services available only through well-defined interfaces reduce coupling.
Components should be interoperable across systems [200]	Interoperability is strongly supported within systems by encapsulation and standard interfaces, and across systems through the Meta-Artifact
The systems development process should produce systems that are reusable, adaptable, and extensible [200]	The Meta-Artifact provides reusability, adaptability, and extensibility
The systems development process should have a balance between product and process [200]	A quality of the Meta-Artifact is that product and process are different views, not different things

More generally, MAP, through the Meta-Artifact, articulates the tacit knowledge embedded in human routines, organizational culture, and existing systems as artifacts (see Appendix B, Meta-Artifact and Tacit Knowledge). Revealing the tacit, unstated, and taken for granted assumptions underlying organizational practices against which organizational knowledge is utilized by the organization's stakeholders on a day-to-day basis has unforeseen consequences, e.g., the risk of calling into question an organization's practices.

Summary

The Meta-Artifact Process (MAP) provides a solution to the constrained problem (see Constrained Problem and Solution Approach for the Constrained Problem sections) through its extensions and enabling technologies.

MAP extensions:

- Meta-Artifact

- Domain Rules Analysis

- Bifurcated Architecture

MAP combines these extensions with the following enabling technologies and methods:

- Object-Oriented Technology (OOT)

- Visual modeling with the Unified Modeling Language (UML)

- COTS integrated tools for modeling, code generation, and database management

- Architecture centricity

- Iterative, incremental methods

The Meta-Artifact is the electronically linked set of all of the artifacts of development, amplified by three extensions of MAP and five enabling technologies and the Meta-Artifact's recursive use in MAP for its own development.

The Meta-Artifact provides a knowledge management narrative across time for the artifacts of development for the domain. The Meta-Artifact continuously supplies the teleological remedy to the entropy that otherwise occurs with time, as development progresses and during operation and maintenance of the system, through its application in the Meta-Artifact Process (MAP), thereby converting knowledge captured in the Meta-Artifact into action and action into additional knowledge in the Meta-Artifact (see Appendix B, Meta-Artifact and Time).

Three of the five enabling technologies, OOT, UML, and integrated tools provide the technology infrastructure for producing and managing the Meta-Artifact. Because of the consistent UML representation and integrated tools, the Meta-Artifact provides the multiple views required by diverse stakeholders (e.g., customers, managers, and auditors) and developer disciplines (e.g., system, software, hardware, database, network, and human factors engineers). The Meta-Artifact facilitates understanding of the system through the active semantic chain created by the electronically linked artifacts. Through its interaction with the enablers and other extensions, the Meta-Artifact represents the totality of the solution space (see Impact of Domain Rules Analysis section) in which the system resides, with several distinguishing properties and qualities useful in development. Iterative, incremental methods assure that the Meta-Artifact represents the totality of the solution space from the outset (see Unified Modeling Language section).

Architecture provides a unifying vision and guiding principles for a system's structural elements and the collaborations among them (see Architecture Centricity section; [19 and 22]). For MAP, the guiding principles would include consistent application of the extensions and enablers (Table 6). MAP preserves the integrity of the architecture and applies it, as represented by the architectural baseline, for the entire lifecycle of the system (see Architecture Centricity section). MAP's architecture centricity strongly supports Domain Rules Analysis by providing a stable baseline as the analysis proceeds over many iterations and increments.

Domain Rules Analysis uses a new concept, domain rules (see Appendix B), to help set the boundaries of the domain and, in conjunction with the volatile variability captured for the Bifurcated Architecture, to keep the focus on the problem domain. Domain Rules Analysis, by assuring a comprehensive analysis of the problem domain before developing a solution, provides the completeness needed for the Meta-Artifact to represent the totality of the solution space, including frameworks, patterns, and components (see Appendix B, Framework and Pattern) from which target systems can be specialized and composed.

The Bifurcated Architecture separates the volatile variability (see Appendix B) – represented by business rules – normally embedded in code, into an external database so that authorized users can directly change it (see Figure 18). The collection of the business rules in an explicit form provides a knowledge management narrative of the controls resulting from the application of MAP. The rules can be maintained more efficiently and accurately than when they are embedded in code; they also become more accessible for such purposes as analysis, audit, and training.

Bifurcated Architecture extends the domain analysis concepts of commonality and variability in two ways:

- Differentiating stable variability from volatile variability (see Appendix B, Variability, Stable Variability, and Volatile Variability)
- Providing physical as well as logical separation (see Bifurcated Architecture to Encompass Commonality and Variability section)

Using Domain Rules Analysis, Bifurcated Architecture explicitly identifies volatile variability and continues to exploit the dichotomy between stable commonality (represented by domain rules) and volatile variability (represented by business rules) through implementation, operation, and maintenance, not just during design and implementation, or only for a particular function, purpose, or aspect. Coupled with Domain Rules Analysis, Bifurcated Architecture provides a means to systematically and completely analyzes a problem domain in a way that identifies its business rules, then to allocate them to an external database, regardless of what the basic function, purpose, or aspect of the resulting components may be.

Chapter 3 includes a detailed procedure for applying MAP.

Chapter 3: Methodology for the Meta-Artifact Process

Overview

The Procedure in this chapter provides step-by-step guidance for applying the three

extensions of MAP: the Meta-Artifact, Domain Rules Analysis, and the Bifurcated

Architecture. The focus of the Procedure is on extracting domain and business rules

through application of Domain Rules Analysis. To achieve this extraction, all sixteen

steps of the Procedure are applied within each iteration, during the requirements and

analysis activities (see Iterative, Incremental Methods section and Figure 6). After they

are extracted and assigned to use cases (see step 5 and Unified Modeling Language

section) and analysis classes (classes that are subject to decomposition into design classes

as development proceeds, e.g., Boundary, Control, and Entity classes [22] ; see step 10),

domain and business rules are treated as any other requirements, so with one exception,

the Procedure does not cover the remaining development activities for the domain and

business rules. The exception is the externalization of the business rules (see step 16),

which leads to the Bifurcated Architecture. The Meta-Artifact, as the integrated

collection of all of the artifacts of development, documents and guides all of the steps of

the procedure (as it does for all MAP development activities of the iterative, incremental

methods on which MAP builds). Each step in the Procedure adds new artifacts to the

Meta-Artifact or modifies existing artifacts, possibly even deleting some. As the steps

are repeated through successive iterations and as artifacts are created from and/or linked to other electronic artifacts within an iteration, the Meta-Artifact is applied recursively. The five enabling technologies (OOT, UML, Integrated Modeling Tools, Architecture Centricity, and Iterative Incremental Methods are in the background, providing the technology infrastructure for the Procedure (see Dynamic Interdependence of the Eight Elements of MAP and Figure 6).

Many existing methodologies provide detailed guidance for developing object-oriented systems, beginning with capturing the requirements in use cases – e.g., the Unified Development Process (USDP [22]) and the Rational Unified Process (RUP [149], a specific, detailed implementation of USDP). Both USDP and RUP are well defined, widely used, with powerful, comprehensive tool support. MAP integrates seamlessly with RUP and USDP because it does not modify any aspects of those methodologies. Rather, MAP uses the standard UML notation to capture the new concepts of domain rules and volatile variability (business rules). Through the Procedure and Domain Rules Analysis, MAP uses a methodology like USDP or RUP to comprehensively analyze and entire domain, rather than the requirements for a single system. In this way, MAP complements existing methodologies, building on them rather than replacing them, eliminating the need to develop new methodologies and train developers in their use. The references below, primarily to USDP, are brief, intended to establish the interface of entry- and exit-points between MAP and existing methodologies, but not to reiterate what the references already explain extremely well. The Procedure only addresses RUP and USDP by reference, as exemplars of existing methodologies that provide a foundation for the Procedure.

The early identification of business rules is valuable for planning. Because the externalized rules co-reside with the components (Figure 18), the distributed physical storage requirements for the externalized rules are available early in development for deployment planning.

The figure and table captions in this and the next two chapters are annotated with the corresponding steps in the Procedure, to illustrate both the procedure and the prototype that is discussed in the next two chapters.

Figure 23 represents the Procedure as a UML activity diagram.

The MAP procedure uses examples from the two domains of interest – Accounting Information Systems (AISs, see step1.1.1) and Command and Control (C2, see step1.1.2). With the wide variability between the two domains, they show – as described in steps 14 and 15 below – how MAP may apply across multiple, diverse domains.

There is a widely accepted body of domain knowledge from which the domain rules (e.g., Table 16 and Table 17) can be extracted for both the AIS and C2S domains [e.g., 54, 89, 135, 142, and 205]. This body of knowledge consists of primary sources, such as text books and field manuals, that incorporate and systematize information from sources that may be more direct, but less accessible, as well as the more direct sources themselves. Such direct sources include natural or legislated laws, standards and regulations promulgated by professional organizations, and works by widely recognized researchers or practitioners in the domain intended primarily for other researchers and practioners. Other mature domains are likely to have similar primary sources of domain information.

Because MAP focuses on the entire domain through Domain Rules Analysis, any established information about the problem domain could supplement the primary sources in defining the domain, as well as determining how to support domain rules extracted from the primary sources. Two secondary sources of domain information (see step 2) would be particularly useful in this regard:

- Traditional subdomains, functions, methods, processes, and procedures
- Established processing cycles, business processes, or patterns

The body of knowledge for the AIS and C2S domains includes such supplemental information (e.g., Table 2, Table 3, Table 18, and Table 19), as described in the following procedure (see step 2).

Other sources to supplement the primary sources may be available, such as Requests for Proposals (RFPs) from external organizations and internal requests for new systems or changes to existing systems (see step 3).

Procedure

The following steps are presented in a logical order for serial application. Unless there is a statement to go back to a previous step for a given condition, the rule is to go to the next step. If there is a need to go back to a previous step n, then the same rule applies after completing step n: proceed to step n + 1. However, there are opportunities for significant parallel and iterative application for teams of developers.

Figure 23 shows the parallelism and iterativeness. Table 15 summarizes the steps to provide an overview of the procedure flow.

1. Identify the domain and extract its domain rules (see Table 3, Table 16, Table 17, Figure 2, Figure 5, and Figure 25)

    1.1. Determine explicit definitions of the domain from primary sources. The definitions often provide implicit domain rules (see step 7) – such as those contained in natural or legislated laws, standards and regulations promulgated by professional organizations, or works by widely recognized researchers or practitioners in the domain. The following are examples:

        1.1.1. AISs monitor an organization's monetary dimension of economic activity by processing data according to known rules and delivering precise information that is useful to those who manage the organization's activities as well as to interested outsiders. The information comprehensively covers revenues, expenses, assets, liabilities, and profits [229, p. 19].

        1.1.2. Command and Control is the exercise of authority and direction by a properly designated commander over assigned and attached forces in the accomplishment of an assigned mission, through planning, directing,

195

coordinating, and controlling the forces in the accomplishment of the
mission. Computers and communications are used to process and transport
information. Intelligence, surveillance, and reconnaissance (see p.36) are
the sources of the data for the information [54, p. 29].

1.2. Extract the domain rules and other requirements from primary sources as part of
the requirements elicitation (e.g., Table 16 and Table 17). The primary sources
provide most if not all of the domain rules, but the supplemental and other
sources (e.g., a user request or a solicitation; step 9) may add detail or even
additional domain rules (see step 7, implicit domain rules).

As with any other requirements when using iterative methods, not all domain
rules must be identified initially. Developers may choose to proceed with an
iteration before all sources are identified or thoroughly examined.

Supplemental sources may reveal embedded domain (implicit) rules (see step 7).
However, it is important to identify the domain rules before moving to the
solution space to get maximum benefit from Domain Rules Analysis. Also, as
shown in Figure 23 , developers may extract domain rules from primary,
supplemental, and other sources in parallel.

Figure 23 also shows the iterative interaction between determining explicit
definitions of the domain and extracting domain rules. The domain definitions
might point to other primary sources of information that would contain domain
rules. Some definitions might even contain explicit or implicit domain rules.

Conversely, primary sources and extracted domain rules might suggest domain definitions or clarify or modify existing definitions.

The possibility of implicit domain rules embedded in explicit domain definitions or supplemental sources would be a case of tacit knowledge that needed to be articulated as artifacts. The knowledge management framework described above (see the Background of the General Problem section above), which examines knowledge through three lenses: breakdowns, narratives, and time, would be especially useful in discovering such implicit domain rules.

The interaction among the domain rules and domain definitions would exploit the iterative methods used by MAP, because as new capabilities (or domain rules) are exposed, developers can choose whether to implement them in the current iteration or a later one (step 8).

Table 15: Summary of MAP Procedure Steps

MAP Procedure Summary
Each of the following steps may add artifacts to the Meta-Artifact or change existing artifacts. As the steps are repeated through successive iterations and as artifacts are created from other electronically linked artifacts within an iteration, the Meta-Artifact is applied recursively. See Figure 6 for how these steps, performed during the requirements and analysis activities of each iteration, relate to the overall flow of MAP and the dynamic interaction of its eight elements. See the narratives for the steps for references.
1. Identify domain and extract its domain rules
1.1 Determine explicit definitions of the domain from primary sources
1.2. Extract the domain rules and other requirements from primary sources as part of the requirements elicitation
1.3 Identify use cases from the domain rules and requirements available at this

MAP Procedure Summary
Each of the following steps may add artifacts to the Meta-Artifact or change existing artifacts. As the steps are repeated through successive iterations and as artifacts are created from other electronically linked artifacts within an iteration, the Meta-Artifact is applied recursively. See Figure 6 for how these steps, performed during the requirements and analysis activities of each iteration, relate to the overall flow of MAP and the dynamic interaction of its eight elements. See the narratives for the steps for references.

point
2. Extract domain rules and other requirements from supplemental sources
2.1 Identify the traditional subdomains, functions (or applications), methods, processes, and procedures and extract domain rules and other requirements
2.2 Identify the established processing cycles, business processes, and patterns and extract domain rules and other requirements
3. Extract domain rules and other requirements from other sources (e.g., from a request or solicitation
4. Identify any requirements from steps 1-3 containing volatile variability (business rules or rules of engagement), to be captured in use cases
5. Allocate requirements from steps 1-4 to use cases – preferably using a standard template to capture the event flows and other details such as non-functional requirements
6. Allocate any non-functional domain rules to all use cases that they affect, along with other non-functional requirements for use-case realization during design
7. Assess completeness by tracing all requirements, including non-functional requirements and volatile variability, to use cases
8. Decide whether to add or modify use cases in this or later builds or iterations. When using USDP or RUP, the decision to defer would be influenced by the current iteration and phase
9. Identify analysis classes and create collaborations to realize the use cases
10. Allocate requirements previously allocated to use cases to analysis classes
11. Allocate any non-functional domain rules to all analysis classes that they affect, along with other non-functional requirements for realization during design
12. Assess completeness by tracing all requirements, including non-functional requirements and volatile variability, to analysis classes
13. Decide whether to add or modify analysis classes in the current or a later build or iteration
14. Assess the analysis classes for commonality.
15. Assess the analysis classes for stable variability
16. Asses analysis classes for volatile variability (business rules) and allocate to an external repository for implementation. Begin next iteration at 1.

Table 16 and Table 17 contain examples of domain rules that are readily available for the AIS and C2S domains.

Table 16: AIS Domain Rules

Domain Rule	Prescriptive Description
Duality	Offset each increment to resources with a corresponding decrement, and vice versa. Characterize increments by transferring in (purchases and cash receipts) and the corresponding decrements by transferring out (sales and cash disbursements) [163, p. 562].
Accounting Equation	Ensure Assets = Liabilities + Owner's Equity
Income Equation	Ensure Revenues – Expenses = Net Income (or Net Loss)
Accounting period	Ensure transaction effective date between Accounting Period End Date and Accounting Period Begin Date
Accrual	Calculate portion of expense or revenue attributable to this accounting period, based on when the corresponding purchase or sale event occurred, not when cash is received or disbursed
Realization	Recognize when the expense occurred based on when the physical item or service was received
Matching	Match revenue that occurred in the accounting period with associated expenses
Money Measurement	Provide a common unit of measure for all calculations by translating all measurements into monetary units
Entity	Define the boundaries of the organization for which accounts are kept and reports are made
Going Concern	Prepare financial reports based on the assumption that the organization will continue its current operations indefinitely, not based on current liquidation value
Cost	Value assets based on original cost, not current value and not adjusted for inflation or deflation (i.e., using only monetary units attributed to the purchase at the time of purchase)
Consistency	Do not change the accounting method for a kind of event or asset from one accounting period to the next in order to enhance

Domain Rule	Prescriptive Description
	comparability of accounting reports from period to period
Conservatism	Recognize revenues and gains slower than expenses and losses
Materiality	Do not measure or record events that are insignificant, applying consistency and conservatism in determining significance [35, pp. 1-8 – 1-11]

(Related to Table 16, see step 1, and extract domain rules as part of the requirements.)

Table 17: Nine Principles of War

Domain Rule	Prescriptive Description
Set the Objective	Direct every military mission toward a clearly defined, decisive, and attainable objective. Commanders direct the use of available combat power toward clearly defined, attainable, and decisive goals. The proper objective ("purpose") in battle is the destruction of the enemy's combat forces. To do this, however, subordinate commanders must be given "terrain objectives" toward which they move.
Take the Offensive	Seize, retain, and exploit the initiative. Offensive action is the most effective and decisive way to attain a clearly defined common objective.
Mass the Effects	Mass the effects of synchronizing the employment of overwhelming combat power at the decisive place and time to gain the objective. Achieve military superiority at the decisive place and time. Mass in this sense does not mean more men. Military superiority can be attained against a more numerical enemy if you have superiority in such things as weapons, leadership, morale, and training. Mass is generally gained by maneuver.
Use Forces Economically	Employ all combat power available in the most effective way possible to gain the objective; allocate essential combat power to secondary efforts. Allocate to secondary efforts minimum essential combat power. This is a misleading term because it does not mean what it sounds like. It does not mean do the job with minimum combat power. Note that the principle pertains to secondary efforts, and it is the means by which a superior general achieves mass as defined above. Mass and Economy of Force are on opposite sides of the same coin.
Maneuver Combat Power	Place the enemy in a position of disadvantage through the flexible application of combat power. Position your military resources to favor the accomplishment of your mission. Maneuver in itself can produce no decisive results, but if properly employed it makes decisive results possible through the application of the principles of the offensive, mass, economy of force, and surprise. It is by maneuver that a superior general defeats a stronger adversary.
Use Unity of Command	Designate a single decision maker responsible for all activities related to an operation. Focus all activity upon a single objective.
Be Secure	Never permit the enemy to acquire an unexpected advantage. Another definition would be to take all measures to prevent surprise. A unit in bivouac, for example, uses outposts and patrols for security.
Use Surprise	Strike the enemy at a time, at a place, or in a manner for which he is unprepared. Accomplish your purpose before the enemy can effectively react. Tactical or strategic surprise does not mean open-

Domain Rule	Prescriptive Description
	mouthed amazement. Thus, a corps may be surprised by an attack it has seen coming for several hours if this attack is too powerful for it to resist by itself and if no other unit is within supporting distance.
Use Simplicity	Prepare clear, uncomplicated plans and clear, concise orders to ensure thorough understanding.

(Related to Table 17, see step 1, Identify the domain and extract its domain rules.)

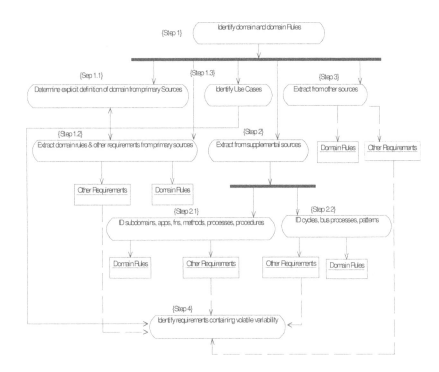

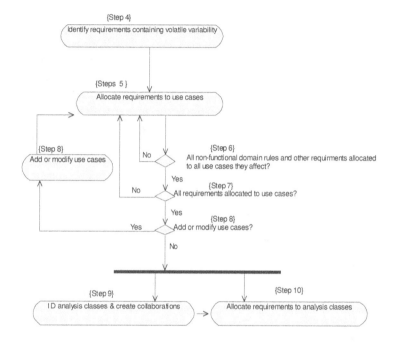

{Step 4}
Identify requirements containing volatile variability

{Steps 5 }
Allocate requirements to use cases

{Step 8}
Add or modify use cases

No

{Step 6}
All non-functional domain rules and other requirments allocated to all use cases they affect?

Yes

No

{Step 7}
All requirements allocated to use cases?

Yes

Yes

{Step 8}
Add or modify use cases?

No

{Step 9}
I D analysis classes & create collaborations

{Step 10}
Allocate requirements to analysis classes

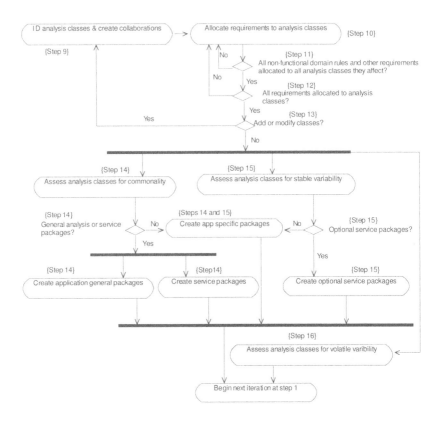

Figure 23: Activity Diagram for Procedure

1.3. Identify use cases from the domain rules and requirements available at this point.

Use cases are identified following standard practices [e.g., 22], but including

domain rules among the requirements from which use cases are inferred. In this

way, domain rules help to identify the initial set of use cases. The process of allocating requirements to use cases involves creating appropriate use cases to capture the requirements: a use case created for one requirement may be an appropriate use case to allocate other requirements to.

Event flows describe use cases and use cases are realized by the collaborations that are suggested by the event flows (see Figure 27 and Figure 35). Developers can infer analysis classes – e.g., Boundary, Control, and Entity classes subject to decomposition as development proceeds – and their relationships from the event flows, which they in turn can use to construct the collaborations.

2. Extract domain rules and other requirements from secondary sources.

   2.1. Identify the traditional subdomains, functions (or applications), methods, processes, and procedures and extract domain rules and other requirements. Using readily available source material for the domain, it is easy to obtain diagrams like Figure 2 [89] and Figure 24 [135, p. 11], showing standard subdomains, applications, and processes. Textual information in [135] and [64] leads to the diagram in Figure 5.

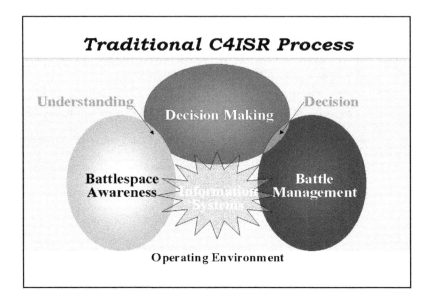

Figure 24: Traditional C4ISR Process

(Related to Figure 24, see Procedure step 2, subdomains and patterns.)

As with domain rules, other requirements from the supplemental sources should be viewed as domain information to avoid imposing a solution before the domain has been analyzed. For example, subdomains should not be equated to subsystems. Subdomains reside in the problem space; subsystems reside in the solution space. Likewise, functions (or applications), methods, processes, and procedures would be defined in terms of the problem space. Not imposing a solution is especially a concern for applications, which are strictly from the solution space. Extracting domain information from applications involves a range of difficulty, depending on such factors as the documentation and the technology used to develop the application.

2.1.1. Roughly speaking, the subdomains (see Table 2, Table 3, and Figure 24) break the domain into more manageable chunks [254] as with any decomposition. The chunks would help organize the functions and applications into related groups, many of which may exist in a mature domain.

2.1.2. In practice, a domain is likely to have many existing computerized applications and other information defining its boundaries. In accounting, existing computerized applications would include general ledger, purchasing, accounts payable, accounts receivable, asset management, and so on. In Command and Control, existing computerized applications would include Global Command and Control System, Contingency Theater Air Planning System, Joint Maritime Command Information System, Maneuver Control System, Advanced Field Artillery Tactical Data System, and Joint Tactical Information Distribution System [54]. Researchers, practitioners, and written material in professional disciplines (e.g., accounting, military command) related to the domain, or documents maintained by the organization, are obvious sources of other information, such as methods (e.g., requisition-based purchasing (compared to a Just-In-Time purchasing) in the AIS domain or Electro-Optical sensing in the Airborne Reconnaissance subdomain of C4ISR), processes (e.g., the workflow for purchase orders in the AIS domain), and procedures (e.g., the approval procedure at certain points in the purchase order flow or the authorization procedure for Air Tasking Orders for Airborne Reconnaissance).

2.1.3. Functions in the context of *Breakdowns* are from the perspective of users of the system in conjunction with the functions they perform, which they want the system to support. This use of the word would be familiar to various stakeholders providing requirements, such as the person whose function was purchasing in a business or airborne reconnaissance in the military. In some cases, functions might be referred to as applications (not necessarily computerized) or roughly correspond to applications (see Figure 2). The related methods, processes, and procedures would then be decompositions of the functions or applications. For example:

Domain: Command and Control (see Figure 5 and Table 2)
Subdomain: Battle Management
Function: Planning
Method: Collaborative
Process: Meeting
Procedure: Electronic dissemination

Domain: AIS (see Figure 2 and Figure 25
Function or Application: Purchasing
Method: Requisition
Process: Electronic workflow
Procedure: Departmental policy

2.2. Identify the established processing cycles (see Table 18, Table 19, and Figure 25) business processes, and patterns and extract domain rules and other requirements. (A business process [22, p. 198] is much broader than the process referred to in step 2; both terms, however, are widely used, so *Breakdowns* includes both senses. Pattern here is used in the sense of a generic collaboration [22, p. 247].)

For mature domains, there will be established processing cycles, business processes, or patterns, such as those in Table 18 and Table 19, which are taken from public sources [e.g., 142 and 238].

The processing cycles and patterns suggest what the functions should accomplish and how they and components within them should interact (collaborate). These cycles and patterns may be viewed as business processes (also referred to as business use cases), each of which may involve multiple use cases [22, p. 198]. Each row in Table 18 and Table 19 represents a collaboration among objects for the realization of the related use cases (references to use-case realizations or realization of use cases are based on the descriptions of such realizations in [22]). Analysis of these collaborations would reveal some of the commonality and variability in the domain, as discussed in steps 14 and 15 below.

Table 18: Standard Accounting Cycles (Step 2.3, standard processing cycles)

Cycle Name	Description
Cash payments	Supplier or vendor invoice, receiving report, written check
Cash receipts	customer checks and remittance advices
Payroll	Time cards, paychecks
Production	Materials requisition, labor time cards, production order, operations list
Facilities	Documents supporting the purchase of property, plant, and equipment
General ledger	Adjusting, closing, and correcting entries and input from various

Cycle Name	Description
	feeder cycles, e.g., expenditure and sales
Financing	Capital raising, e.g., bank notes, bond agreements, common stock issuances
Investment	Stocks, bonds, CDs, repurchase agreements
Purchasing	Purchase requisition, purchase order
Sales	Customer order, customer purchase order, bill of lading invoice

Table 19: Basic C2 Patterns [238] (Step 2.3, determine existing patterns)

Name	Description
Plan	Translation of higher Commander's vision / intent into specific Courses Of Action (COAs) in a compressed plan cycle for preparation and execution by subordinate elements. Define battle space areas of operation for synchronization and control. Generate alternate courses of action and evaluate against most likely and dangerous adversary actions. Develop synchronized schedule of tasks and activities for subordinates to prepare and execute. Develop integrated, combined effect operations plan to include all the battlefield functional areas.
Prepare	Activities by the unit before executing, to improve its ability to conduct the planned operation, including plan refinement, force protection, rehearsal, reconnaissance, integration and coordination of warriors and resources, inspections, and movement to planned locations.
Execute	Apply combat power to accomplish the planned mission, exercise control through assessment of battlespace to understand the situation in order to make execution and adjustment decisions for battle management.
Assess	Monitor and evaluate on a continuous basis throughout planning, preparation and execution the current situation and progress of an operation and the evaluation of it against criteria of success to make decisions and adjustments.

3.   Extract domain rules and other requirements from other sources (e.g., internal

requests or RFPs)

4. Identify any requirements from steps 1-3 containing volatile variability (business rules or rules of engagement), to be captured in use cases (step 5) and allocated to analysis classes (step 10) for implementation in an external repository (step 16)

5. Allocate requirements from steps 1-4 to use cases [22: chapter 7, section 13.4.1 and section 14.3.1] – preferably using a standard template to capture the event flows (see Figure 35) and other details such as non-functional requirements [22, pp. 138-139].

   To graphically allocate domain rules, non-functional requirements, or volatile variability to use cases, use one of the standard UML notation mechanisms for recording special information, e.g. a tagged value, constraint (free-form text or formal syntax such as Object Constraint Language), or documentation notation (see Figure 25; [22, pp. 158, 203, and 211; 21, pp. 77-83]).

6. Allocate any non-functional domain rules to all use cases that they affect, along with other non-functional requirements for use-case realization during design. Domain rules or other requirements may be non-functional – system properties not directly implemented by a specific function [22, pp. 66, 116, and128-129]. As system properties, a non-functional requirement (including non-functional domain rules) may affect multiple use cases. E.g., a performance requirement should be allocated to each use case whose functional requirements would affect that performance criterion. If non-functional domain rules and other requirements are allocated to all use cases they affect, go to step 7; otherwise, go to back to step 5.

7. Assess completeness by tracing [22, p. 39] all requirements, including non-functional requirements and volatile variability, to use cases. Determine if all functional and non-functional domain rules have been captured and allocated to appropriate use cases [22, pp. 158, 186, 189-190, 203, 206, 211]. If requirements from the supplemental sources (i.e., capabilities identified from supplemental sources, see step 1.2 above), seem to go beyond the domain rules, then determine if implicit domain rules are embedded in the capabilities, or if the capabilities are unnecessary, e.g., they lie outside the domain, they are redundant, or they are obsolete (especially likely from sources in step 2). If all requirements are allocated to use cases, go to step 8; otherwise, go back to step 5.

8. Decide whether to add or modify use cases in this or later builds or iterations. When using USDP or RUP, the decision to defer would be influenced by the current iteration and phase. There may be only one iteration [148], especially in the first phase (called Inception in RUP). While not all use cases are usually identified during the Inception phase, only architecturally significant use cases are needed to establish the scope and assess a candidate architecture for the system [22, pp 346-348]. When the use cases seem reasonably complete for the iteration, proceed with the analysis. If use cases need to be added or modified, go back to step 5.

9. Identify analysis classes (Entity, Boundary, or Control classes [22]) and create collaborations to realize the use cases (see Figure 26; [22, pp. 204-206]). Use event flows from use cases (see steps 1.3 and 5) to infer analysis classes. Automatically

generate an executable from the sequence diagram to verify and validate the
interactions of the analysis classes

Care should be taken to not realize use cases for requirements from other sources
(step 3) before those for steps 1 and 2, to avoid moving too quickly to the solution
space. Other sources would include those from a request to produce a system,
ranging from a formal competitive solicitation from an organization separate from
that of the developer, to an informal request from an internal customer. Such
requests typically initiate development of new systems. The requirements provided
in this way, by their nature, tend to focus only on a single application, or at most a
product line, a family of applications [67], or (for C2Ss) a system of systems in a
domain. Such requirements are often solution oriented. By focusing on the entire
problem-space rather than solutions for a requested system – even when the request
involves multiple applications, such as a product line – Domain Rules Analysis
considers requirements that are often not contained in a request.

10. Allocate requirements previously allocated to use cases to analysis classes [22, pp.
181-182]. The details for doing this can be obtained from RUP, USDP, or
comparable sources. According to the USDP, this would be done largely as part of
the analysis activity [22, p. 203]. Depending on the extent of the changes, add the
analysis classes in another build [148] in this iteration or wait till the next iteration.
This is a good example of how the iterative methods complement Domain Rules
Analysis.

To graphically allocate domain rules, non-functional requirements, or volatile variability to analysis classes, use one of the standard UML notation mechanisms described in step 5 (see Figure 26).

11. Allocate any non-functional (see step 6) domain rules to all analysis classes that they affect (see Figure 26), along with other non-functional requirements for realization during design. If non-functional domain rules and other requirements are not allocated to all analysis classes they affect, go back to step 10.

12. Assess completeness by tracing all requirements, including non-functional requirements and volatile variability, to analysis classes. If all requirements are not allocated to analysis classes, go back to step 10.

13. Decide whether to add or modify analysis classes in the current or a later build or iteration (see step 8 re decision to defer). If analysis classes need to be added, go back to step 9.

14. Assess the analysis classes for commonality (the USDP analysis method includes a means of identifying commonality). Allocate common analysis classes to application general or service packages as discussed below in this step. If no commonality is identified, allocate the analysis classes to application specific classes as stable variability [22, p. 202]. The USDP analysis method relies on the collaboration analysis done for use-case realizations. These collaborations reveal common analysis classes, that is, classes that appear in multiple use cases [22, p. 45]. Such classes perform responsibilities (which become operations in the design process) in each of the collaborations in which they appear. The responsibilities a class performs in a

collaboration constitute its role in that collaboration. Such classes are often domain or business entity classes that represent long-lived or persistent information [22, p. 184].

Developers may capture such commonality by placing the common analysis classes in general-analysis packages to show in the analysis model how multiple application-specific packages depend on the application-general package [22, p. 202]. Developers can then use general analysis packages as the basis for a layered architecture, with an application-general layer below an application-specific layer [22, p. 73]. During design, the analysis packages may become one or more subsystems.

The dependencies between an application-general package and one or more application-specific packages are the result of associations among classes in those packages [22, p.212]. These associations may in turn lead to generic collaborations [22, p. 247].

Developers may also capture commonality by identifying a service package for services provided by functionally related analysis classes [22, pp. 199-200]. Classes are functionally related when:

- A change in one is very likely to require a change in the others
- Removing one makes the others function incorrectly or inadequately
- They interact heavily with each other

Service packages may be separate or contained within application specific or application general packages. They may apply to the entire domain [22, p. 192]. When service packages capture stable variability (see step 15) they represent options, which may be mutually exclusive or provided in combination [22, p. 200].

There are also many generic mechanisms (22, pp. 75, 246-249, 370) such as transaction management that would be required by multiple applications in the domain of interest and other domains, but existing to support rather than embody the domain rules. These functions are above the system-software (e.g., operating system, language translator, database management [21, p. 401], or networking software) and middleware layers (e.g., CORBA, GUI builders), but they are independent of the business context [72, p. 2], or not business-specific [22, p.73]. While generic mechanisms are not of direct interest for Domain Rules Analysis, they offer important opportunities for reuse, reliability, and interoperability, in that they are designed to realize reusable collaborations [22, pp. 247-249].

Domain rules assist in identifying commonality for applications in the domain. By definition, if an application is in the domain, at least one of the domain rules applies to it, so any application in the domain would have to contain the class or classes capturing at least one domain rule. Any classes corresponding to domain rules used in more than one application would be common classes, possibly participating in generic mechanisms or collaborations for reuse. Analysts could use this to greatest advantage by cross-referencing domain rules and analysis classes to identify candidates for common classes, generic collaborations, and generic mechanisms early in the analysis process.

15. Assess the analysis classes for stable variability. Allocate analysis classes representing stable variability to optional service packages (see step 14; [22, p. 201]) or application specific packages [22, p. 202]. In addition to this analysis-level assessment, variability may also be identified through subsequent analysis or design of the common analysis classes. The common responsibilities may be used to define a base class, with the roles for different collaborations and aggregations accomplished by derived classes (see Figure 67 and Figure 68).

Classes derived from such base classes would identify stable variability (see step 4 and Domain Rules section) among applications within the domain, because they would be based on the domain rules, traditional functions, and processes for the domain, or because they were not subsequently identified as volatile in step 16. For example, the component to match depreciation expense with sales for a period would not be the same as the component to match direct materials costs with sales, but the portions of each component related to ensuring that appropriate components were invoked to match sales and expense for the accounting period (i.e., the domain rule for *Matching* in Table 16) would be inherited. Furthermore, the specialized aspects of the two would seldom require change. There are basic depreciation methods that seldom change. What changes is which method to apply to a class of assets, which depends on the business rules (see steps 4 and 16).

16. Assess analysis classes for volatile variability. Allocate volatile variability

 (business rules) to an external repository for implementation (see Figure 18, Figure

 50, and Figure 51), rather than program code (see pp. 137, 157, and 259). The

 externalized rules would be the mechanism for implementing the prescriptive

 instructions of the controls that enforce the rules (see p. 158, Table 20 and Table 21);

 the repository or repositories containing them would be assigned to the components

 whose business rules they would enforce (see p. 281) and would physically reside in

 the same processor and memory as the components to which they were assigned (see

 Figure 18). Components could be comprised of multiple active classes, so the

 database of rules for a component might support multiple active classes   The details

 for the externalization are covered in Chapter 4 and Chapter 5.

Table 20 and Table 21 show some typical business rules, along with their

prescriptive instructions, for the two domains. The Rules of Engagement (ROE) in

Table 21 are taken from [239].

Table 20: Sample Business Rules for AIS

Business Rules	Prescriptive Instructions
Financial – Compliance with budget policies	If upon approval of this request for purchase order, total encumbered dollars for this subsidiary ledger account would be greater than the budget for that account, reject the request
Operational – Compliance with authorization polices	If the amount for this purchase order exceeds the signature authority of the Buyer (purchasing agent), reject the purchase order
Regulatory – Compliance with tax law and regulation	if the type of asset-type specified for this subsidiary ledger account does not match the asset type for this depreciation method, reject the transaction: either the wrong subsidiary

Business Rules	Prescriptive Instructions
	ledger account is being used to set up this asset or the wrong depreciation method has been specified
Fraud – Compliance with legal and policy requirements	Select all transactions for a specified subsidiary ledger account for a specified time period exceeding a specified dollar amount, then process the details of those transactions (e.g., name of vendor, name of purchasing agent, address of vendor, shipping address) through specified neural network to detect patterns of fraudulent activity.

(Related to Table 20, see steps 4 and 16 identify and allocate volatile variability (business rules).)

Table 21: Sample Rules of Engagement

Rules of Engagement	Prescriptive Instructions
Use armed force as the last resort	When possible, the enemy will be warned first and allowed to surrender. Armed civilians will be engaged only in self-defense. Civilian aircraft will not be engaged without approval from above division level unless it is in self-defense.
Avoid harming civilians unless necessary to save US lives.	If possible, try to arrange for the evacuation of civilians prior to any US attack. If civilians are in the area, do not use artillery, mortars, armed helicopters, AC-130s, tube- or rocket-launched weapons, or M551 main guns against known or suspected targets without the permission of a ground maneuver commander, LTC or higher (for any of these weapons). If civilians are in the area, all air attacks must be controlled by a FAC or FO. If civilians are in the area, close air support (CAS), white phosphorus, and incendiary weapons are prohibited without approval from above division level. If civilians are in the area, do not shoot except at known enemy locations. If civilians are not in the area, you can shoot at suspected enemy locations.
Avoid harming	Public works such as power stations, water treatment plants, dams,

Rules of Engagement	Prescriptive Instructions
civilian property unless necessary to save US lives	or other utilities may not be engaged without approval from above division level. Hospitals, churches, shrines, schools, museums, and any other historical or cultural site will not be engaged except in self-defense.
Treat all civilians and their property with respect and dignity.	Before using privately owned property, check to see if any publicly owned property can substitute. No requisitioning of civilian property without permission of a company-level commander and without giving a receipt. If an ordering officer can contract for the property, then do not requisition it. No looting. Do not kick down doors unless necessary. Do not sleep in their houses. If you must sleep in privately owned buildings, have an ordering officer contract for it.
Control civilians engaged in looting	Senior person in charge may order warning shots. Use minimum force but not deadly force to detain looters. Defend Panamanian (and other) lives with minimum force including deadly force when necessary.
Secure and protect roadblocks, checkpoints, and defensive positions	Mark all perimeter barriers, wires, and limits. Erect warning signs. Establish second positions to hastily block those fleeing. Senior person in charge may order warning shots to deter breach. Control exfiltrating civilians with minimum force necessary. Use force necessary to disarm exfiltrating military and paramilitary. Attack to disable, not destroy, all vehicles attempting to breach or flee. Vehicle that returns or initiates fire is hostile. Fire to destroy hostile force. Vehicle that persists in breach attempt is presumed hostile. Fire to destroy hostile force. Vehicle that persists in flight after a blocking attempt IAW instruction 2b is presumed hostile. Fire to destroy hostile force.

(Related to Table 21, see steps 4 and 16 identify and allocate volatile variability

(business rules).)

Modeling Considerations in Applying the Procedure

For distributed applications, sequence diagrams are especially useful for specifying interprocess communications [21, chapter 18]. The lifelines for each process (or passive classes used by the processes) then document what is to be accomplished within the classes (active classes for processes) on the sequence diagram. Both actions and states can be shown visually and documented on the lifelines when using the IBM Rational Rose Technical Developer (RRT) tool. When generating code, state machines ultimately implement the states and actions on these lifelines. Since code is not generated for the business rules, it is at this point that the design work for business rules would diverge from that for program code.

Sequence diagrams would remain useful for business rules, both for analyzing and for documenting them. Design and subsequent development workflows within each iteration would differ for business rules in that they would be transformed into production rules rather than program code, then maintained in a business rules database (see pp. 137, 157, and 259; [126]). While state machines could still be useful to document the detailed logic within states and actions on a lifeline, activity diagrams would be more suitable, since the nature of such logic focuses on the activities that take place within an object rather than event-ordered behavior of an object [21, chapter 21].

Allocating the detailed logic to the states within a class provides a powerful organizing technique, at an even finer granularity than a class. The logic for an individual state can be highly cohesive as well as providing smaller chunks for understandability.

Representing a class visually down to the granularity of individual states, combined with the externalized business rules, enhances understandability by leaving only a relatively small amount of total system logic to be realized in program code. As part of the active semantic chain, the enhanced understandability provided by the finer level of visual granularity when representing a class with state diagrams contributes significantly to the Meta-Artifact's quality of promoting understanding (see p. 187).

The next chapter describes a prototype built with this procedure, to demonstrate the application of the procedure described in this chapter. Figures representing various electronic artifacts of the prototype are annotated with the step or steps of the procedure involved in creating the artifact.

Chapter 4: Prototype Description

Overview

Starting with the AIS domain represented in the use case diagram in Figure 25, the

prototype is described in a series of UML diagrams. The context of the AIS domain is

shown in Figure 2.

Figure 25 represents with use cases the standard accounting cycles as described in a

modern accounting textbook [142]. There is a one-to-one relationship between the

accounting cycles in the textbook and the use cases, requiring no application of MAP.

UML notation is used for the accounting cycle use cases to correspond to the notation

used in developing the prototype. The prototype is limited to the purchasing cycle of the

AIS Domain to have a reasonable scope for the level of detail required for an executable

prototype. As noted in, the AIS and C2S domains have common control needs. Other

cycles in the AIS (see Figure 25) would use many of the same kinds of controls as those

included in the prototype (see Table 22)

Description

Developers working with stakeholders who had appropriate domain expertise would have to realize [22] each of the use cases in Figure 25 to develop a complete AIS. Although the prototype realizes only one of the ten use cases, development of the prototype and

analysis of the remaining use cases leads to the pattern in Figure 26. This pattern for AIS

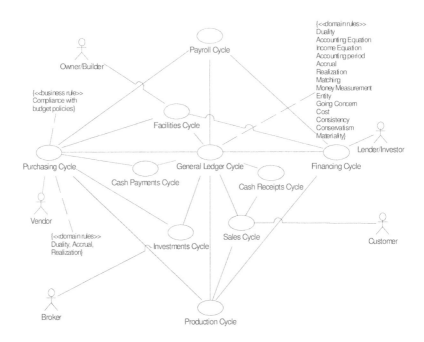

Figure 25: Use Case Diagram of AIS Domain

(See Procedure, step 5, for allocation of domain rules and other requirements to use cases.)

use-case realizations (references to use-case realizations or realization of use cases are based on the descriptions of such realizations in [22]) could be used at least as a starting point for the remaining use cases in Figure 25. The basic pattern is to accept input and display results for stakeholders using the system, parse the input, validate input that was successfully parsed, then post it to the General Ledger.

Table 22: Common AIS Controls

Accounting Cycle	Type of Control							
	$ Range	Time Period	Author ization	Vendor	Product	Dept	Buyer	Cstmr
Cash Payments	X	X	X	X	X	X	X	
Cash Receipts	X	X			X	X		X
Facilities	X	X	X	X		X	X	
Financing	X	X	X			X		
General Ledger	X	X	X	X	X	X	X	X
Investments	X	X	X	X	X	X	X	
Payroll	X	X	X			X		
Production	X	X	X		X	X		X
Purchasing	X	X	X	X	X	X	X	
Sales	X	X	X		X	X		X

Each use case would have additional persistent data to record. In the case of Purchasing, the prototype is limited to showing the application of the Meta-Artifact Process (MAP).

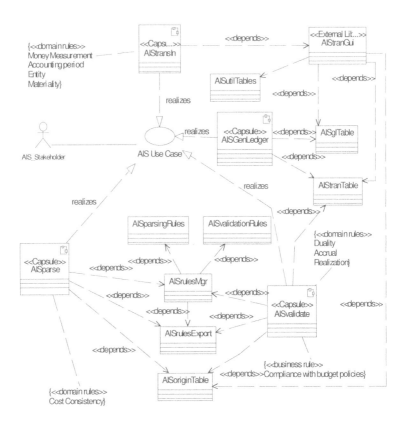

Figure 26: AIS Use Case Pattern Realization

Figure 26 shows Domain and Business Rules applicable to AIS pattern, as constraints in braces; see Procedure step 10.)

The externalization of business rules – a defining characteristic of the Bifurcated Architecture of MAP – is largely an implementation issue, so is included in the prototype.

Other aspects of implementing the Purchasing use case require only well established data processing practices, such as those for maintaining persistent data or printing documents and reports. The prototype uses posting the General Ledger as an example of such maintenance, because that would occur in all of the use cases.

Another implementation feature of the prototype is its distributed, concurrent, real-time architecture. Such an architecture simplifies composition of target systems [93, 95] because it emphasizes the OOT characteristic of encapsulation and requires coupling only through published interfaces. The architecture also simplifies scalability through concurrent or simultaneous processing of transactions (see p.257). This directly demonstrates the use of two of MAP's enabling technologies, OOT and integrated tools. The key role of the integrated tools here is to automatically generate code for the publish and subscribe interfaces. The use of UML is demonstrated indirectly in the implementation when the visual model is animated, showing behavior within and among the implemented components. Such animation provides an example of Visual Verification and Validation (see Animated, rather than static, views of development artifacts section).

The use-case realization for Purchasing in Figure 27 is an instance of the AIS use case pattern realization (the prefix PO is from Purchase Order; see Figure 67 and pp. 283 and 360 for other transaction types and naming conventions). The active classes (with the stereotype <<Capsule>>) are related to the Purchasing use case with realization arrows.

Active classes are related to passive (boundary or control [22]) classes or data (entity

[22]) classes with dependency arrows. Other detail is elided from this diagram to keep

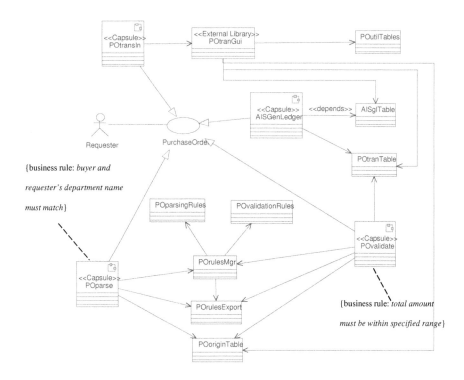

Figure 27: Purchasing Use-Case Realization

the focus on the major active classes and their dependencies that are needed to carry out the flow of events in Figure 35.

Other relationships among the classes are shown in the collaboration, sequence, and state diagrams, e.g., Figure 28, Figure 29, and Figure 30, which continue the realization of the use case, ultimately through the code, e.g., Figure 31.

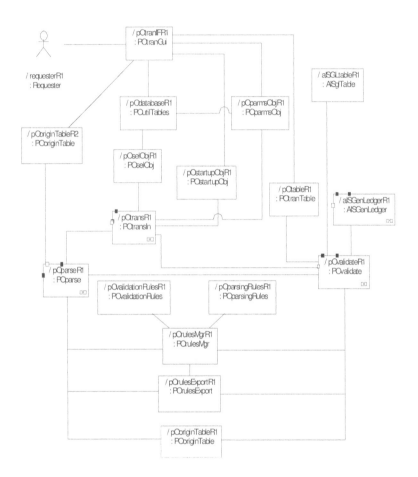

Figure 28: Purchasing Collaboration

(Related to Figure 28, see Procedure step 9, create collaborations, to realize the

Purchasing use case.)

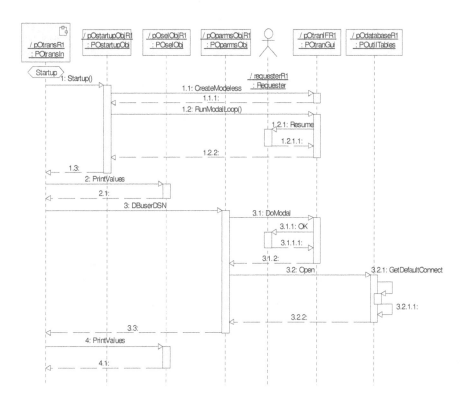

Figure 29: Purchasing Startup Sequence – Part 1

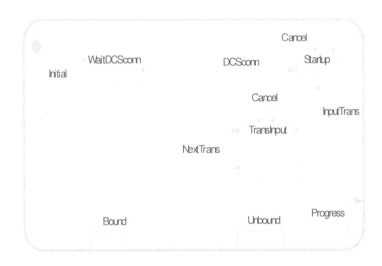

Figure 30: State Diagram for POtransIn Active Class (Capsule)

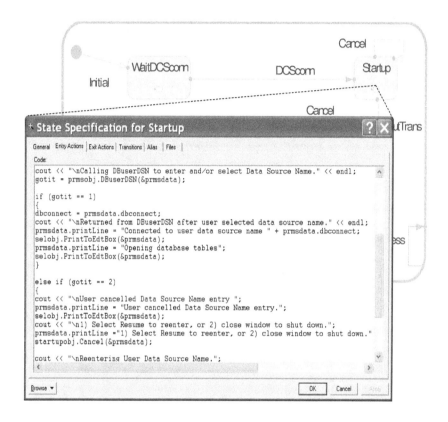

Figure 31: Code in State Diagram for POtransIn

The classes in the use-case realization diagram Figure 26 are analysis classes that may be decomposed into design classes, as in Figure 27, Figure 68. Figure 68 decompose the validation class for Purchasing as a hierarchy, but the decomposition could be into classes that were not so related.

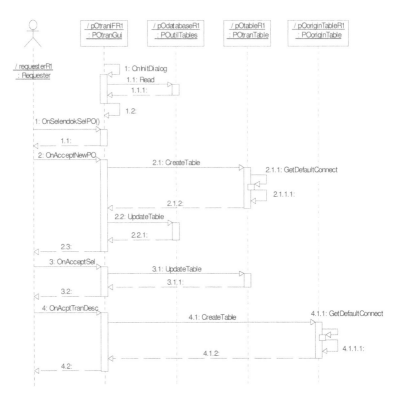

Figure 32: Purchasing Input Sequence

The collaboration in Figure 28 shows the relationships among all of the classes in Figure

27, rather than the relationship of the active classes to the use case and their

dependencies. Figure 28: Purchasing Collaboration does not show the content or

sequence of any messages. This allows the collaboration to represent multiple scenarios or message sequences, as shown in Figure 48, Figure 36, Figure 32, Figure 49, and Figure 33.

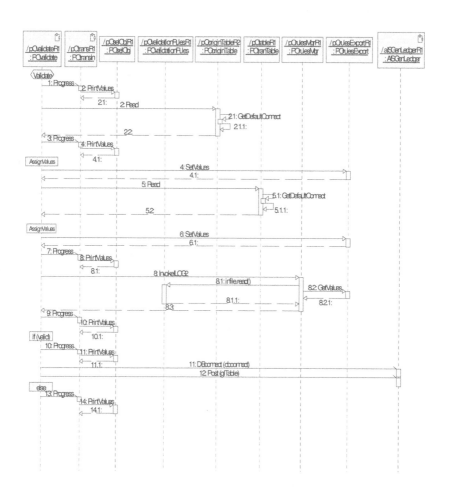

Figure 33: Purchasing Validation Sequence

The integrated tools provide an option to select one or more of the sequence diagrams to overlay onto the collaboration diagram, as in Figure 34. The class names that appear on Figure 27 are automatically propagated by the tools through the other diagrams, a key enabler for the Meta-Artifact (see Table 8 and p.171).

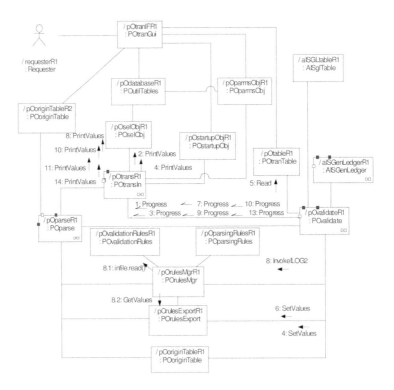

Figure 34: Purchasing Collaboration with Sequence Overlay

(Related to Figure 34, see Procedure step 9, create collaborations, to realize Purchasing use case. As noted in Chapter 3, the Procedure does not cover design, so the sequence overlay is not mentioned in the Procedure.)

The lifelines beneath the active classes in the sequence diagrams represent the processing that must be done by them. As much detail as desired can be placed on these lifelines, including states and local actions (see TransInput state and AssignValues local action on the lifeline underneath POtransIn on Figure 36).

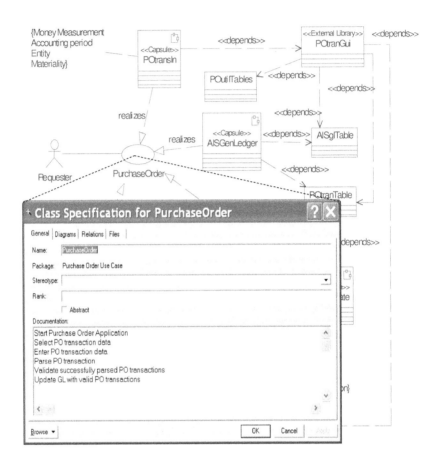

Figure 35: Purchasing Flow of Events

(Related to Figure 35, see Procedure step 9, using the flow of events to infer analysis classes.)

The messages on the message lines are the actual messages sent and received through the publish and subscribe interfaces of the active classes.

Likewise, the call messages from active to passive classes are actual operations of the passive classes.

For the first iteration of a sequence diagram, it may be more efficient to enter names on the message lines that capture the sense of what is being accomplished, rather than trying to determine message names and related interface names suitable for implementation at first, especially if capturing domain expert input in a group setting. Later, analysts can systematically arrive at global naming conventions, using the descriptive names captured during the sessions with domain experts.

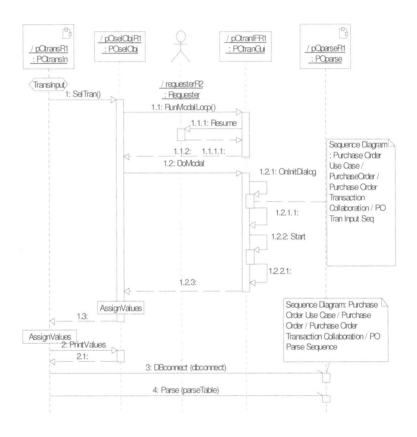

Figure 36: Purchasing Startup Sequence – Part 2

The integrated tools will then create an executable from the sequence diagram. Code can also be entered directly in the states or transitions. The individual states and transitions within a class provide a graphical means of organizing processing steps into small, highly cohesive groups, greatly easing their comprehension and management. In this sense, state diagrams extend the concept of encapsulation, allowing the structure – represented by the states and transitions – within a class to remain stable even when code is changed. Existing code, such as algorithms, can be pasted into the states and transitions, or in the case of C++, used through include statements or library files. The tools provide integrated code generation and management regardless of the source, through the visual model. This integrated management enables multiple features of the Meta-Artifact (see Table 8). For C++, code is generated in standard .cpp and .h files, as in Figure 37.

```
// {{{RME classifier 'Logical View::AIS Purchase Order Tran::POtransIn'
#if defined( PRAGMA ) && ! defined( PRAGMA_IMPLEMENTED )
#pragma implementation "POtransIn.h"
#endif
#include <RTSystem/POtransIn.h>
#include <POtransIn.h>
#include <POtranGui.h>
extern const RTActorClass RTDBase_Agent;
INLINE_METHODS void POtransIn_Actor::enter4_Startup( void )
{
        // {{{USR
        ParmsObj prmsobj;
        ParmsData prmsdata;
        StartupObj startupobj;
        SelObj selobj;
        int cancelSwitch = 0;
        int gotit = 0;
        cout << "\nCalling Startup to create modeless dialog box for instructions to user." << endl;
        prmsdata.printLine = "Select Resume, then acccept or change User Data Source Name, then 1) select OK, or 2) Select Cancel.";
        startupobj.Startup(&prmsdata);
        cout << "\nReturned from Startup after creaing modeless dialog box." << endl;
        prmsdata.printLine = "Starting up Purchase Order Transaction.";
        selobj.PrintToEdtBox(&prmsdata);
        cout << "\nCalling DBuserDSN to enter and/or select Data Source Name." << endl;
        gotit = prmsobj.DBuserDSN(&prmsdata);
        if (gotit == 1)
        {
        dbconnect = prmsdata.dbconnect;
        cout << "\nReturned from DBuserDSN after user selected data source name." << endl;
        prmsdata.printLine = "Connected to user data source name " + prmsdata.dbconnect;
        selobj.PrintToEdtBox(&prmsdata);
        prmsdata.printLine = "Opening database tables";
        selobj.PrintToEdtBox(&prmsdata);
        }
        else if (gotit == 2)
```

Figure 37: Example of Code Generated for POtransIn.cpp

Chapter 5: Prototype Execution

Overview

This chapter describes in detail the execution of the prototype. The prototype provides a series of screens to instruct the user in the operation of the prototype. The user selects a database, then enters data on a transaction screen. When the user accepts the transaction, processing begins. The prototype displays the results of the processing, then prompts the user for another transaction.

Input

Figure 48 shows the first part of the startup sequence. The requester selects a database through the Human Machine Interface (HMI) provided by POtransGui (Figure 38, Figure 39, and Figure 40). This selection invokes a method that provides the location of the database, using the standard Open Database Connectivity (ODBC) services.

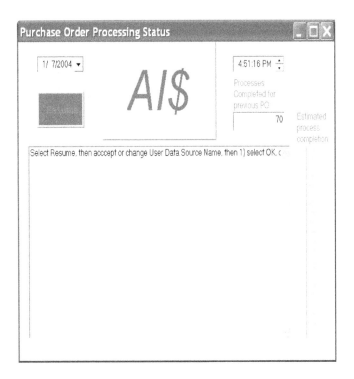

Figure 38: Requester Startup Interaction

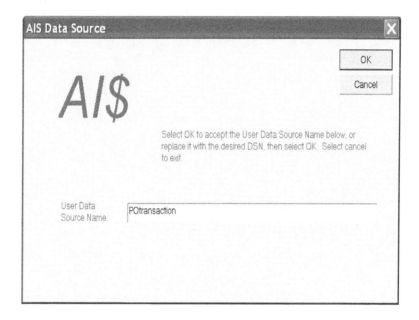

Figure 39: Database Selection

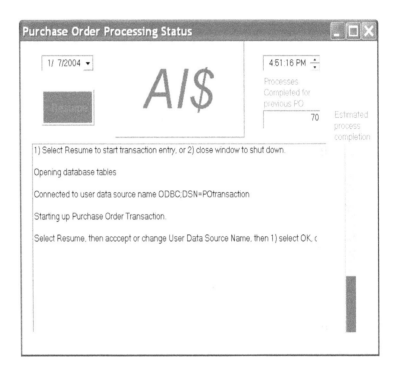

Figure 40: Requester Startup Interaction – Opening Database Tables

Figure 36: Purchasing Startup Sequence – Part 2 shows the interaction between the

Requester actor and the HMI to begin input, once startup is completed. The sequence

diagram in Figure 36 shows a link to the actual input sequence in Figure 32. Both the startup sequence and the input sequence use database tables (POutilTables in Figure 48),

to fill the drop down lists and record the input.

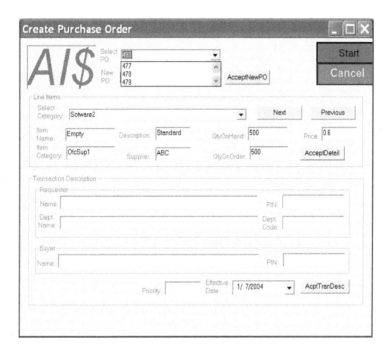

Figure 41: Drop Down List for Selecting Existing Purchase Order

The Requester can select an existing Purchase Order from the drop down list (Figure 41)

loaded from a utility table or create a new Purchase Order (Figure 42). The user indicates

the final choice of Purchase Orders by selecting the AcceptNewPO button. Next, the requester selects a category from another drop down list loaded from another utility table

(Figure 42).

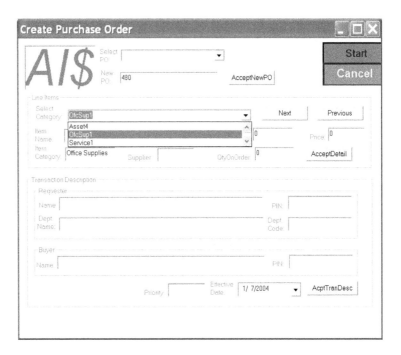

Figure 42: Requester Creates a New Purchase Order

The Requester displays the line items from the category by selecting the Next or Previous

buttons (Figure 43). To add the displayed item to the Purchase Order, the Requester selects AcceptDetail, for any number of items (Figure 43). The accepted items are

written to the transaction detail table (POtranTable, Figure 52)

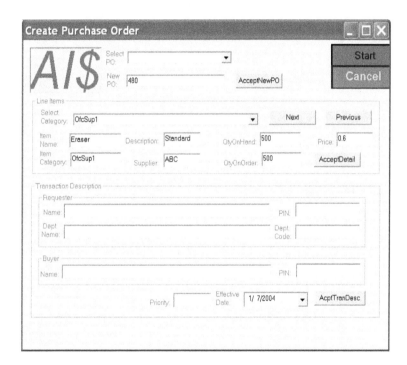

Figure 43: Select Line Items

The Requester next describes the transaction by entering the Requester and Buyer Names

and related information, with the priority, for transaction authorization (in addition to the

251

access authorization provided by the operating system to start the application) and subsequent processing. In operation, the application would route the document in Figure 44 to the buyer, who would enter his PIN, but for demonstration purposes, the PIN can be entered by the requester.

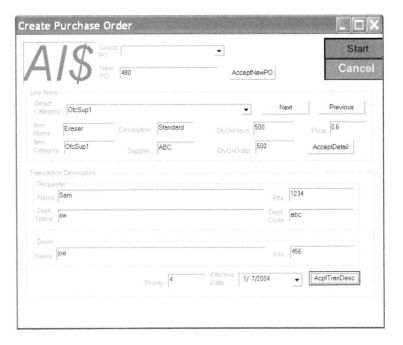

Figure 44: Enter Transaction Description

The application automatically fills Effective Date box with the system date. If the

transaction effective date is not the current date, the Requester can select the drop-down

calendar (Figure 45) to select another date (which would be parsed to ensure that it was in

Figure 45: Date Selection

the correct accounting period). When finished (e.g., after reviewing or correcting the information), the requester selects the AcptTranDesc button. This writes the transaction description detail to a database table, POoriginTable (Figure 49).

The drop-down calendar, along with all of the other features of the HMI, are provided by the standard services of Microsoft Visual Studio. This is an example of the full integration of the modeling tools with the C++ development environment, which includes the ODBC database support. All of these integrated services are managed through the modeling tool, providing an example of how the integrated tools are a major enabler of the Meta-Artifact.

During processing, the other active classes (POparse, POvalidate, and AISGenLedger in Figure 49, Figure 52, and Table 15) send messages to POtransIn (the active class that controls the HMI) to display the status of the transaction (Figure 46 and Figure 47).

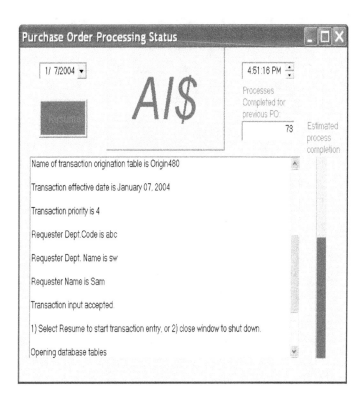

Figure 46: Message Display – Part 1

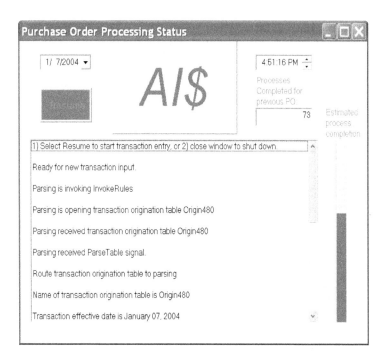

Figure 47: Message Display – Part 2

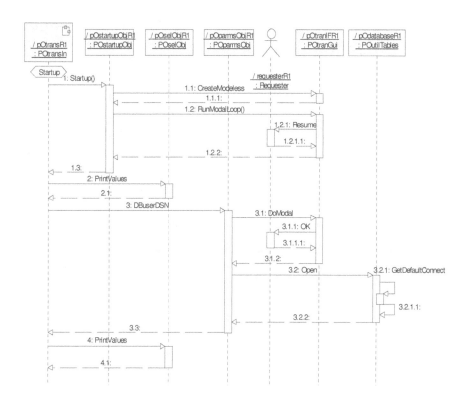

Figure 48: Purchasing Startup Sequence – Part 1

Processing

POtransIn sends the name of one of the tables created by the input sequence,

POoriginTable (Figure 49), to the component POparse, via the publish and subscribe

interfaces (see p. 130 for discussion of role of publish and subscribe interfaces in analysis). POparse can be located anywhere on the network, as can the database table.

POparse assumes control when it receives the signal that passes the name of the table to

be parsed. Up to that time, POtransIn had controlled the processing flow. It has invoked

operations on passive classes (e.g., POstartupObj and POselObj in Figure 48), which in

turn invoke operations on POtranGui, a Dynamic Link Library component that controls

the HMI. POtranGui is an example of a boundary (or interface) class [22]. POtranGui

"invokes" an operation on the Requester actor, in that the actor must perform some action

for processing to resume.

After it sends the table name to POparse, POtransIn transitions to its state for accepting

input for another transaction concurrently (or simultaneously, if it were running in a

separate processor) with the parsing of the first transaction. Figure 49 shows the

processing sequence for parsing the transaction.

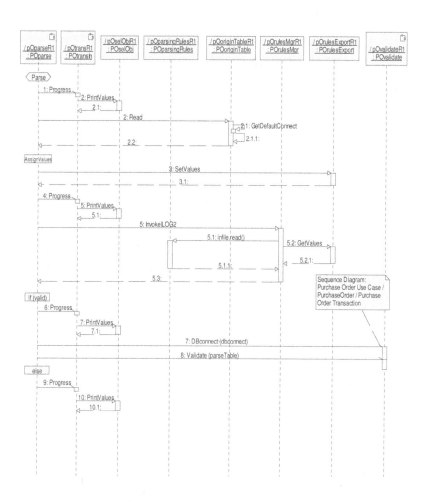

Figure 49: Purchasing Parsing Sequence

Parsing checks the syntax and construction of the transaction. Simple input tests, such as whether the value entered in a field is of the right type or within the right range, can be done as part of the field definition and applied as the Requester enters the value. As an example of what might be done by a parser in this context, the Requester's department name is compared to the name of the buyer. One of the business rules in this case is that buyers are assigned to departments, for control purposes (Figure 50 and Figure 51). This comparison is done by the POrulesMgr (see Figure 49), using business rules stored in the external database named POparsingRules (see Appendix C for implementation details and p. 137 and p. 157) for role of the external database in analysis).

The graphics for figures 50-51 and 53-66 are taken from ILOG Rule Builder, a product of ILOG S.A.

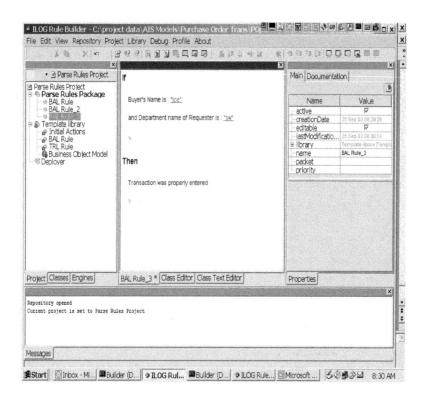

Figure 50: Example of Business Rule in External Database for Parsing

Figure 51: Example of Business Rule in External Database for Validation

While this is an example of a simple business rule, complex rules can also be built by

end-users, as discussed below (see p. 263). Authorized stakeholders such as the

Requester's supervisor would use the drag-and-drop HMI (see e.g., Figure 50, Figure 51, and Appendix C) provided by POrulesMgr to maintain POparsingRules.

If the transaction successfully passes parsing, POparse sends the name of the table to POvalidate via the publish and subscribe interfaces and informs the Requester through POtransIn, which invokes the HMI to display a message that the transaction passed parsing. POvalidate can also be located anywhere on the network, as can the database table that passes parsing. If the transaction fails parsing, POparse informs the Requester through POtransIn, which invokes the HMI to display a message that the transaction failed parsing.

POvalidate assumes control when it receives the signal from POparse that passes the name of the table to be validated. As with POtransIn, after sending the table name to POvalidate, POparse transitions to its state for receiving the next table name for processing and can process concurrently (or simultaneously, if it is running on a separate processor) with POvalidate and POtransIn. POvalidate processes two tables for validation; it extracts the name of the second table from the table name that POparse sent.

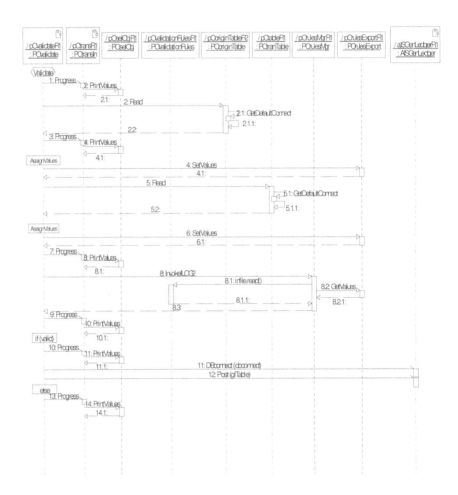

Figure 52: Validation Sequence

Figure 52 shows the processing sequence for validating a transaction. Once the syntax

and construction of the transaction had been accepted by parsing, validation would apply more complex business rules. The example in the prototype is to first obtain the buyer's name from the table received from POparse, then determine the name of the transaction line-item table from the name of the parsed table. POvalidate sums the amounts of the line items and tests whether the total is within the authority of the buyer (Figure 53). If the transaction passes validation, POvalidate sends it to AISGenLedger for posting.

Creation and Maintenance of Rules

Complex business rules for both parsing and validation could be entered by end-users with the same drag and drop interface. One such rule was implemented for parsing, where the external rule invokes an operation in POparse that reads a database table of valid buyer and requester department combinations. This rule is included with the rule described above. POrulesMgr is an inference engine that will test all rules in working memory for a match. If POrulesMgr found no match for the simple rule, but found (by executing the operation in POparse) the combination of buyer and requester department name in the table of valid combinations, it would find the transaction to be properly entered.

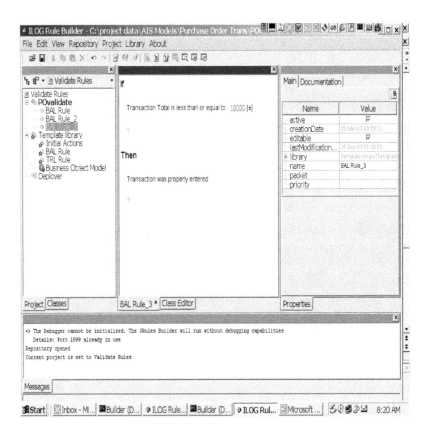

Figure 53: Validation of Transaction Total

These two examples for parsing illustrate how the COTS product manages the external

business rules. Attributes (e.g., the name of the buyer) and operations (e.g., the operation to process the table of valid combinations) are exported via .cpp and .h files

(POrulesExport in Figure 49) for use by the inference engine (POrulesMgr). The HMI in

the COTS product makes these attributes and operations available to the user in drop-

down lists for building rules.

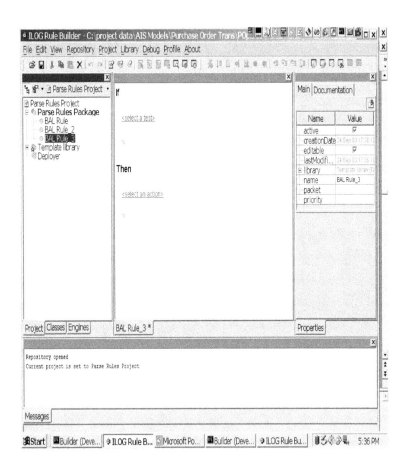

Figure 54: Example of Empty Rule Template

Figure 54 shows an empty template for building a new rule. The next step is to select an

attribute or operation for the test (Figure 55 and Figure 56).

Figure 55: Select an Attribute

Figure 56: Attribute Selected

Once this is done, the user selects a relational operator (Figure 57).

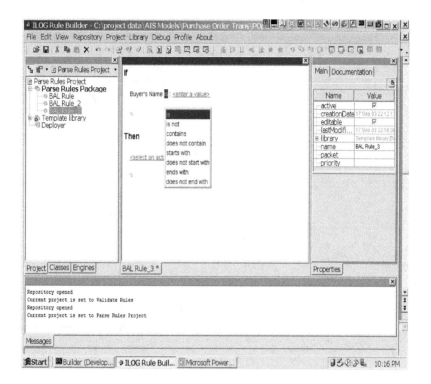

Figure 57: Select Relational Operator

After choosing a relational operator, the user may choose to enter either a literal value or

to select another attribute or operation (Figure 58, Figure 59, and Figure 60).

Figure 58: Choose to Enter a Literal Value

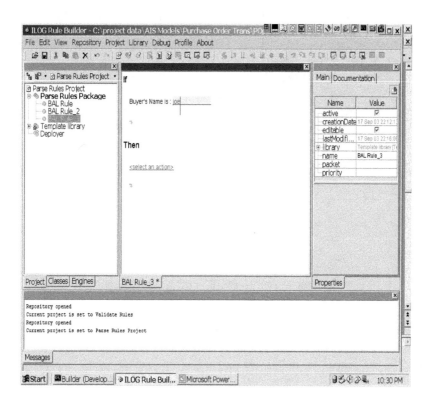

Figure 59: Value Entered

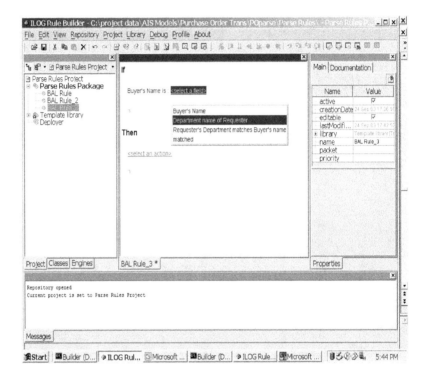

Figure 60: Select an Attribute

For a simple rule, this would complete the condition (if part) part of the rule. The last step would then be to select an operation to execute for the action (then part) if the condition were met (Figure 61).

Figure 61: Select an Operation to Execute for the Action

For complex rules, the user could select one or more additional logical operators (Figure

62) to construct complex conditions prior to specifying an action (Figure 63, Figure 64,

Figure 62: Select Logical Operator

and Figure 65).

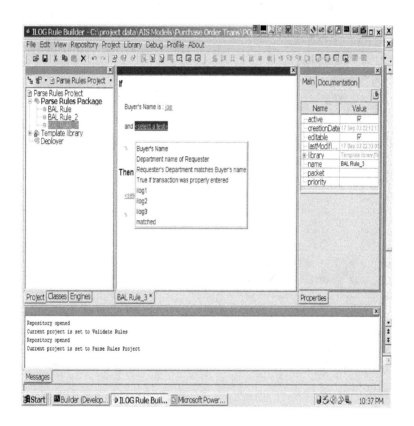

Figure 63: Select Additional Attribute

Figure 64: Select Additional Relational Operator

Figure 65: Enter Additional Value to Match

In all of the figures referenced for building rules, the attributes and operations have all

been identified with natural language descriptions. The HMI of the COTS product

provides a table (Figure 66) where developers and users can enter such descriptions for

each of the exported attributes and operations. This decouples any naming conventions developers find useful for their purposes from descriptions that are meaningful to the end

users.

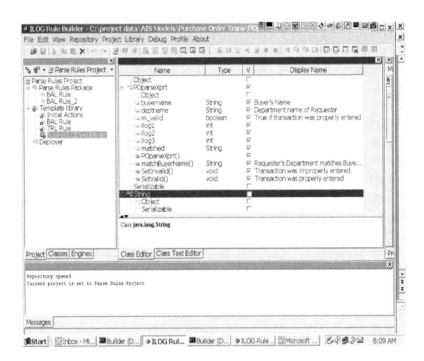

Figure 66: Enter Natural Language Descriptions

The COTS product uses type information for the exported attributes to reduce user errors

in building rules. For example, the relational operators in the drop-down menus are

different, depending on the type of the attribute (e.g., Figure 53 and Figure 56). The

COTS product will also allow objects of classes defined in the application to be exported as attributes, retaining the class of the object as the type.

Inference Engine Considerations

Because it uses a general-purpose inference engine, the COTS product is capable of rules processing that goes beyond what is needed for this prototype. An expert system is not needed for the prototype. A custom component that would build complex if-then statements from attributes, operations, and logical or relational operators entered into a table by end-users would suffice. The COTS product was selected because it could perform this function as a subset of its capabilities and it had a robust HMI suitable for end-users, without requiring any development. Such availability of a ready-made solution not only reduced the effort in developing the prototype, but also was important in demonstrating the general applicability of the Bifurcated Architecture, both in terms of feasibility and practicality. The COTS product is sufficiently general and robust to satisfy the needs of a range of real-world applications.

Nevertheless, using an inference engine for processing business rules raises some interesting questions. For example, does the prototype have an intelligent agent or multi-agent architecture? As noted, the processing flow is controlled by the active classes. The inference engine, POrulesMgr, is not an active class. Its methods are invoked through an object embedded in the active class using it. Both POparse and POvalidate use it in this manner. This allows the external business rule repositories to be limited to only those rules needed by the active class and to be distributed with the active class. Such cohesion and distribution of business rule repositories simplifies development and maintenance by including only those rules in a database needed by the active class that uses the database. It reduces risk by isolating the direct impact of errors in the rules to the responsible active class, improving reliability. This isolation supports a fundamental control procedure, separation of duties. Users authorized to maintain rules in one database could be physically prevented from changing rules in another database. Cohesion and separation can be further improved within the same database by using multiple packages for rules (see Figure 73 and Figure 74). The COTS product provides robust security management to provide logical access control to repositories and packages.

The POrulesMgr used by POparse is a different instance than the POrulesMgr used by POvalidate (they appear as one instance in the diagrams to simplify the diagrams). Each is controlled by its respective active class, so neither instance of POrulesMgr is autonomous, nor do they learn (adapt), which precludes their being intelligent agents by most definitions [28]. The instances of POrulesMgr do not collaborate directly, so they do not function as a multi-agent system [28 and161], by definition. The prototype would be a rule-constrained, rather than a rule-based, system [99]. In general, the Bifurcated Architecture would apply to such rule-constrained systems. However, because ontologies and taxonomies can be generated from the Meta-Artifact, there is a potential for using the Meta-Artifact to assist in building multi-agent systems (see Generation of Ontologies and Taxonomies section).

Because it is not using an intelligent-agent or multi-agent architecture, the prototype, and

the Bifurcated Architecture it implements, do not raise any issues of reliability or

complexity beyond what is normal for if-then rules in applications without a Bifurcated

Architecture. The only difference is that the rules can be changed by end users. This

does eliminate some high-risk activities encountered in maintaining such applications that

do not have a bifurcated architecture. First, code is not modified, eliminating the errors

of such modifications (see p. 155 for quantification). Second, the software engineering

process for modifying code is avoided, eliminating substantial costs and delays (see p.

155 for quantification). The end-users might get the logic of their if-then rules wrong,

but they are the same domain and subject matter experts who would provide the logic to

software engineers, which would propagate the same errors in the modified code.

Another question relates to whether the embedding of an expert system in an application

like this prototype might offer some potential use not realized by the prototype (see p.

151). Such potential uses would be beyond the scope of *Breakdowns*, but there are some

that would be worth exploring in future research, including:

- Applications where the distributed processes reside in autonomous platforms
  - o Unmanned military vehicles
  - o Factory robots
  - o Software embedded in civilian, service-oriented products, from toys to medical devices
- Decision Support

- o Standalone

- o Embedded in processes

- Application of artificial intelligence to controls, such as fraud prevention

Use of Polymorphism and Inheritance

POvalidate1 uses polymorphism with the embedded object to invoke the methods of the inference engine. POvalidate1 is a derived class of AISvalidate, as are CDvalidate and ROvalidate (Figure 67). AISvalidate aggregates a class that declares a pure virtual function (see p. 162 for role of pure virtual functions in analysis) named InvokeRules (). POvalidate aggregates a passive class that defines InvokeRules (), which invokes the inference engine object, POrulesMgr.

This polymorphism would be used by all of the special control classes shown in Figure 68; this is an example of how polymorphism promotes extensibility, as discussed in [61]. That is, POvalidate1 would use whichever special controls it needed by aggregating some combination of special control classes (e.g., POvalFin). Each of these special control classes would define its specialized InvokeRules () method. The specialized InvokeRules () methods might invoke different inference engine objects (in order to use different rules repositories), as needed.

Appendix D discusses the rationale for the aggregation shown in Figure 68. While always using the same name for the operation, InvokeRules (), each aggregated class (each of the control classes in Figure 68 might be aggregated with a sibling of POvalidate) would use its own instance of POrulesMgr. Each of the aggregated validation classes is passive class so they could be aggregated as needed. That is, certain transactions would not require all of the validations, such as fraud prevention (Figure 68). Others might require multiple instances of a validation component, such as regulatory compliance, because of the relatively lengthy processing such compliance can entail.

Such polymorphism simplifies the detail coding, because the statement calling InvokeRules () is inherited by each derived class and does not need to be modified. Other benefits of inheritance can be optimized by naming conventions to take advantage of the tool support. For example, interfaces for the active classes are embedded (composite) objects, so if an interface is inherited, it will have the same name as the one in the parent class. By using generic names in the parent class, e.g., valSub for the subscribe interface for validation, the name does not need to be changed for the derived class. The tool has a drop-down list of defined interface classes to easily change the class of which the interface is an instance. Since it is the class that defines the signals (operation names) that an interface can use, the interface on the derived class may access different signals

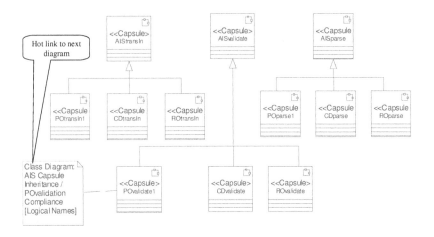

Figure 67: Inheritance Structure for Active Classes

(Related to Figure 67, see step 15 of Procedure in Chapter 3, assess analysis classes for stable variability.)

than the one with the same name on the parent class. In addition, the tools automatically update the trigger information, part of which is the interface name, for transitions when interfaces are updated in this way.

Another significant savings from the use of inheritance is that the API for the COTS

product is inherited and managed by the tools. Developers working with the existing

derived classes or deriving new derived classes do not generally need to be aware of the

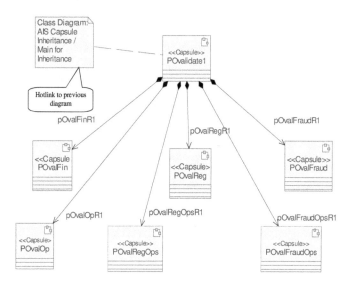

Figure 68: Aggregation of Controls from Figure 91

(Related to Figure 68, see step 15 of Procedure in Chapter 3, assess analysis classes for

stable variability.)

API, as long as the necessary library and include files are available. Developers need

only provide any code that may be needed in addition to what is inherited for the standard

InvokeRules () operation.

The graphical state diagrams are also inherited, offering additional savings in deriving new classes. This is because the first two or three states and their transitions in an active class tend to be very similar, e.g., registering the interfaces with the locator service and waiting for publish and subscribe interfaces to be bound. By using generic names for the states and transitions, few or no changes are needed. If the interface names are preserved, as noted above, even the inherited interface registration statements do not need to be changed. The configuration of the interfaces for the selected protocol, including with an interface is a publisher or subscriber is inherited as well. Finally, the services specified for the publish and subscribe infrastructure are also inherited, which, combined with the other interface details that are inherited as just described, makes the derived class ready to interact in a network environment (or as a separate process in a single processor). Such inherited capability for network communications saves substantial effort and avoids the related risks of such manual effort.

Chapter 6: Research Results, Contributions, and Conclusions

Overview

This chapter presents the research results in terms of the two methods of validation used: the Prototype (described in detail in Chapter 4 and Chapter 5) and the Evaluation (see Validation and Research Approach sections). The sections on Future Research Directions precede the Conclusions and Contributions.

Results

Prototype

The prototype was developed by applying MAP to one of the use cases, the Purchasing Cycle, of the AIS domain (see Figure 25). As a complete application of MAP, the prototype represents MAP for validation purposes. In particular, the procedure for MAP (see Procedure), which applies Domain Rules Analysis (see Definition of Domain Rules Analysis), was used to iteratively refine the Purchasing use case, adding artifacts to the Meta-Artifact in each iteration. An executable prototype, with a Bifurcated Architecture, was generated from the Meta-Artifact. The prototype executes all of the functions (e.g., input, parsing and validation using rules stored in an external repository, database update) described in Chapter 4 and Chapter 5 successfully and very quickly, demonstrating the properties and qualities of the Meta-Artifact (see the Properties section and the Qualities section). The first column of Table 23 lists the properties and the first column of Table 24 list the qualities of the Meta-Artifact. The second column of each table describes how the prototype demonstrates each property or quality. The prototype demonstrates the properties and qualities so completely that the second column of each table is very close to the respective definitions. The specific examples of how the prototype demonstrates the properties and qualities of the Meta-Artifact are shown in Chapter 4 and Chapter 5, by the figures showing artifacts (including generated code for the Prescriptive property) and the related tables and narrative.

Table 23: Prototype demonstration of MAP properties

Property	Prototype provision of property
Current	The artifacts (including those shown in Chapter 4 and Chapter 5) are part of the Meta-Artifact stored in a model element database that is created and maintained by integrated modeling tools. The executing code for the demonstration is generated from the Meta-Artifact
Dynamic	The Meta-Artifact for the prototype is the running system and provides views of the actual system, including all of its artifacts. The integrated tools allow multiple views of the prototype system, not just prescribed static views.
Prescriptive	The Meta-Artifact for the prototype generates the prescriptive code for the executing prototype.
Unifying	The executable components of the prototype have a common user interface, shared databases, and direct interaction with each other. The use of integrated modeling tools based on UML to recursively apply the Meta-Artifact ensures that artifacts of the Meta-Artifact for the prototype have a common representation, a common architecture, and are accessible to all stakeholders
Seen from multiple views with a common representation	The Meta-Artifact of the prototype is represented in UML (as are the representative artifacts in Chapter 4and Chapter 5). UML assures a common representation, with multiple views provided by the integrated tools.
Recursive	The sequence of artifacts presented in Chapter 4 and Chapter 5, shows the recursive application of the Meta-Artifact as artifacts are generated from each other at all levels – including the executable components – with a single architecture, which is captured in the common framework and patterns for the generated executable components.

Table 24: Prototype demonstration of MAP qualities

Quality	Prototype provision of quality
Stakeholder access to current views appropriate to their interest	This quality is provided by the Current, Dynamic, and Multiple-View properties of the prototype's Meta-Artifact
Animated, rather than static, views of development artifacts	The Dynamic and Multiple-View properties of the prototype's Meta-Artifact are used by the integrated tools to provide this quality, e.g., animated views of the visual model
Special-purpose views	The Dynamic and Multiple-View properties of the prototype's Meta-Artifact are used by the integrated tools to provide this quality, e.g., the 26 products defined by DoDAF
Process and product are different views rather than separate things	All activities performed on the actively linked artifacts of the Meta-Artifact for the prototype led directly to the objective system, through MAP's recursive use of the Meta-Artifact to automatically generate the prototype system. The building of the Meta-Artifact in this recursive way provided a systematic process, without separate overhead activities, e.g., to prepare documentation and specifications.
Active semantic chain	All artifacts of the Meta-Artifact for the prototype are actively linked and accessible through the integrated modeling tools

The prototype also satisfies the conditions (provides the capabilities) of the Formal

Research Hypothesis as shown in Table 25.

Table 25: Formal Research Hypothesis Conditions Satisfied by Prototype

Formal Research Hypothesis Capability	Prototype Views Satisfying Capability
1. A capability to comprehensively articulate stakeholders' explicit and tacit knowledge of a domain as artifacts and a means to store the	The prototype was designed beginning with artifacts for the AIS context (see Figure 2), domain rules to cover the domain (Table 16), and use cases for the entire domain (see

Formal Research Hypothesis Capability	Prototype Views Satisfying Capability
artifacts as information that can be converted back to explicit knowledge, to assure adequate understanding and proper implementation of control needs for the domain during the complete lifecycle of systems in the domain	Figure 25). These artifacts capture the high-level semantics of the domain to support the initial realization of use cases into analysis classes and related artifacts and for subsequent enhancement and maintenance activities.
2. A capability to convert the information representing the artifacts of the domain into explicit knowledge to assure adequate understanding and proper implementation of control needs for the domain during the complete lifecycle of systems in the domain	The artifacts, beginning with the initial ones for satisfying capability 1 and continuing with the artifacts in Chapter 4 and Chapter 5, are captured electronically as the Meta-Artifact in the database of actively linked artifacts maintained by the integrated modeling tools used to create the artifacts. The tacit and explicit knowledge captured for the prototype is available through the active semantic chain, which stakeholders can use through the tools to obtain views that provide the explicit knowledge they need at any time during the prototype's life.
3. A capability to use the explicit knowledge of capability 2 to take account of existing systems and potential future systems in the domain during the complete lifecycle of the system under development, because of both the explicit and tacit knowledge contained in other systems on which the system under development may depend, especially when they must interact with each other	The interactions of all use cases for the AIS domain are represented in Figure 25. By applying the domain rules from Table 16 (see Figure 26) when realizing the Purchasing use case (Figure 26-Figure 36), the realization and implementation of the Purchasing use case inherently take account of other applications in the domain.
4. A capability to produce an architecture that reduces or eliminates the disconnects (gaps) among development disciplines and development phases	The prototype, by using the MAP Procedure, allows the same team (consisting of representatives of each discipline) to apply all work flows during each iteration avoiding handoffs from discipline to discipline and phase to phase (see Role of the Meta-Artifact in MAP

Formal Research Hypothesis Capability	Prototype Views Satisfying Capability
5. An architecture that reduces the need to change software source code for maintenance resulting from breakdowns or changing control needs, with resulting reductions in costs and defects	The prototype implements the Bifurcated Architecture using a COTS inference engine to manage the prescriptive rules implementing the controls identified as business rules during analysis and design (see Figure 26 and Figure 27). Controls can be changed directly by end users by making changes to the rules read as data by the inference engine (see Chapter 5 and Appendix C). Adding and changing rules was straightforward and easily verified, validating the basic premise of the Bifurcated Architecture.

The prototype validated the predictive [46] aspect of the research (see Research Approach section) by including structures and patterns (Chapter 4 and Chapter 5) that satisfied the requirements of the domain. The structures in Figure 67 and Figure 68 provide a framework for the Purchasing and Cash Payments (which would include the RO – for Received Order) use cases in Figure 25. Figure 34, with its related sequence and state diagrams (using inheritance and polymorphism as discussed in the Use of Polymorphism and Inheritance section) and the related inherited interfaces provide the corresponding patterns.

Evaluation

The evaluation part of the validation [222] applies *Six Sigma* [see Appendix B; 184] to

the use of one of MAP's extensions [see Table 6], Domain Rules Analysis. Six sigma is

a comprehensive and flexible system for process improvement. The evaluation was

based on a project whose purpose was to develop a safety-critical system for which the

requirements were to be met by applying assertions to the top-level requirements and

allocating them through successive levels of analysis and design to the implementation of

the system. The project is ITAR restricted so identifying details have been redacted, but

are available to authorized persons on request. It will be referred to simply as the Project

below.

The evaluation was done by a team of four systems engineers (including the writer)

working on the Project and a Six Sigma expert as a coach to ensure the proper and

objective application of Six Sigma. The evaluation was done over a period of five

months. All four engineers had received prior training in the use of Six Sigma. The

purpose of the Project was to:

> ... devise an improved process for systematically defining the assertions within a system, to increase the reliability of safety-critical systems. Existing processes are inadequate in their ability to provide assertions that sufficiently cover the required domain.
>
> The assertions of a system are claims, or statements that define the system's proper behavior. For a software system, the assertions may be checked during system operation to verify that the system is functioning properly.
>
> Our ... customer (along with other customers) has expressed the need for software that is more reliable, particularly software used in safety-critical systems .... In summary, it is our customer's position that:
>
> o The software used in software-intensive safety-critical systems is insufficiently reliable.

- o It has been postulated, outside the scope of this Project, that having a sufficiently complete set of assertions can significantly improve the reliability of the software, when used to tailor software development and system monitoring.
  - Assertions offer a cross-checking "mechanism" which when used with test strategies and requirements, provide the ability to detect and avoid defects or incorrect behavior.
  - The use of assertions allows the thoroughness of the specification to be assessed by actively monitoring the real-time system for assessment of proper behavior, particularly, detecting safety-hazard criteria and safety-critical situations.
  - The assertions will aid in the safety accreditation of the system.
  - The assertions will aid in bringing out defects earlier, at a higher level of abstraction.

- o There is no process available to reliably identify a sufficiently complete set of assertions needed to produce software for safety-critical systems.

Key factors here are: (1) how to generate a sufficiently complete set of assertions, (2) how to do it reliably, and (3) how to measure an improvement, relative to the existing approach for generating assertions.

Specifically, this Project [was to] define and apply an improved process for determining a sufficiently complete set of assertions needed to ensure proper operation of the safety-critical … [software].

The team proposed two candidates for the assertion-generation process that attempt to implement the desired properties specified in the Characterize section. These methods are referred to as Meta-Artifact Process/Assertion (MAP/A) and the Assertion Flowdown Process (AFP). The MAP/A process is an extension of the Meta-Artifact Process (MAP), tailored to manage assertions. The basis of the Assertion-Flowdown Process (AFP) is to identify and organize the assertions via an analogy to the formal requirements flowdown process of [structured analysis]. Each process provides an independent point-of-view for the generation of assertions for a specified domain.

MAP/A provides a systematic way of identifying a sufficiently-complete set of high-level assertions, corresponding to domain rules. The object-oriented, architecture-centric approach of MAP then inherently organizes the assertions around the architecture by allocating them to analysis classes, used to realize the domain use cases. Continuing the object-oriented approach into design would decompose the analysis classes and their allocated assertions into design classes and assertions, including hierarchies of design classes and assertions [29].

AFP contrasts with MAP/A by functionally decomposing the high-level assertions – identified through the application of MAP/A – associated with the architecture, rather than continuing the object-oriented approach [29].

The results of the evaluation are summarized in Table 3. The team used a variety of standard Six Sigma tools [29], supplemented with the Scientific/Interpretivist techniques [46] of field study, questionnaire-based survey, and interview-based survey. The evaluation itself was a field study. The object of the field study, the application of Six Sigma to a selected portion of a system under development for the Project, was directly observed by the team. The scoring summarized in Table 3 was a questionnaire-based survey, which collected written data from team members for the scoring matrix. The results were then used for an interview-based survey to record the oral assessments of each team member in order to reach a consensus among the members. A subject matter expert, not a member of the team, reviewed the results as a means of validating the team's conclusions [29]. The Six Sigma project could be repeated by a new team by following the steps in [29], with some adjustment to the area of the system to be studied.

> The two processes were scored against each metric using a scale of 1 to 10, with higher scores indicating greater success. The individual scores for each metric were then combined using a weighting representing that metric's relative importance. Finally, the process achieving the higher total score is declared the better process [29].

MAP/A received an overall score of 165; AFP received an overall score of 142.

In the metrics where the contributions of MAP (via MAP/A) were directly tested – Completeness and Repeatability – the score for MAP was substantially higher than that for AFP. On Completeness, MAP/A scored 60% higher; on Repeatability, MAP/A scored 40% higher. Both of these metrics depended on domain rules, one of MAP's extensions. Given the limited nature of the evaluation – to identify an improved process for identifying assertions – the other two extensions to MAP (the Meta-Artifact and

Bifurcated Architecture) were not assessed. However, the procedure [detailed in the

Procedure section] for MAP was successfully applied to identify a pattern for volatile

variability in the context of assertions for eventual development of a Bifurcated

Architecture (see Figure 69).

Table 26: Comparison of MAP/A and AFP for Generating Assertions

Metric	Description	Discussion	Weight	MAP/A		AFP	
				Score	Weighted Score	Score	Weighted Score
Completeness	To what extent does the process generate a com-plate set of assertions and how well is the degree of completeness determined	Both approaches provide a systematic method of identifying the assertions and are likely to generate assertion sets with considerable completeness. However MAP/A provides a built-in mechanism for assessing the measure of completeness	10	8	80	5	50
Repeatability	To what extent would different users generate the same set of assertions	The MAP/A approach should be somewhat more repeatable since in MAP/A the generation of the assertions is more systematically guided by a set of underlying high-level domain rules.	5	7	35	5	25
Efficiency	Which process can more quickly generate an assertion list	The two approaches are judged equal in efficiency since both require domain experts to identify appropriate assertions for each domain area.	3	5	15	5	15
Implement-ability	What is the effort required to implement the process	Although both approaches require the availability of domain experts, the MAP/A approach additionally requires a facilitator familiar with the principles of Object-Oriented Technology.	4	5	20	8	32
Usability	How difficult will it be to train people to use the processes	There are probably only minor differences in the training requirements. Both approaches require guiding a domain expert into thinking about the right technical areas. The MAP/A approach will require a thorough review of the OO-based MAP/A process whereas the Flowdown approach of AFP only requires familiarity with standard systems-engineering flowdown techniques.	5	3	15	4	20
Total Score					165		142

Since achieving sufficient-completeness was the primary motivation for this [Six Sigma] project, it [was] given a weight of 10. ... MAP/A ... is given the higher score of 8 because it provides for better organization of the assertions and contains a built-in mechanism [domain rules] for measuring the achieved degree of completeness. Since AFP lacks any such mechanism, it only receives a score of 5.

The Repeatability metric addresses the extent to which two different users would arrive at the same answer using the candidate process. A weighting of 5 is assigned to this metric because, although this attribute is less important than completeness, it is still an important consideration and repeatability builds confidence in its use. MAP/A is scored 7 vs. a score of 5 for AFP because assertion development and organization in MAP/A is more systematically guided by the underlying, structured, high-level domain rules.

The Efficiency metric measures how quickly a process can generate a set of assertions. This is given a weight of 3 because it was the collective judgment of the team that efficiency was simply not nearly as important as the technical quality and completeness of the assertions and because none of the evaluated processes takes very long to execute once they have been implemented. Both processes are scored 5 in this area since both are judged adequate in speed.

Implementability is the measure of the effort required to implement the process. This is assigned a weight of 4 reflecting its moderate importance; if implementation is difficult, then adaptation and use of the new technique will be slow. Both processes require the ready availability of domain experts to supply the necessary background knowledge, but MAP/A requires that the users be acquainted with the principles of object oriented technology. Therefore, AFP is given a score of 8 while MAP/A is given a score of 5.

Usability is a measure of the ease in training new users. A weighting of 5 is assigned reflecting the importance of training. AFP has a small advantage in usability since it requires the domain experts to have familiarity only with standard systems-engineering flowdown techniques, while MAP/A requires them to be familiar with the new Object-Oriented Technology. Therefore, MAP/A and AFP scores are 3 and 4 respectively.

The weighted scores for each metric are given in the two shaded columns on the right side of [Table 26] and summed at the bottom. MAP/A achieves a total score of 165 compared to a total score of 142 for AFP. This comparison indicates that MAP/A is the better of the two candidates [29].

The discussion for the Efficiency metric in Table 3 does not explain that the high-level assertions used by both MAP/A and AFP were identified using MAP/A, because AFP lacked the systematic procedure for identifying domain rules, from which the high-level assertions were derived. All assertions (redacted) used by the (Six Sigma) project were identified in a four-hour workshop applying the MAP Procedure to previously defined domain rules, whereas the original process over a period of months did not identify any such high-level assertions. Subsequent decomposition of assertions was then done by both MAP/A and AFP.

The domain rules from which the assertions were derived were themselves identified through eight hours of analysis by one team member, followed by two hours of review with the remaining team members. The ability to identify at least a preliminary set of domain rules quickly for a third domain – represented by the Project, in addition to the C2S and AIS domains examined in [Constrained Problem] – provided further support for the general applicability of the domain rule concept. The domain rules provided the team members with a better definition of the Project's domain than had been previously available and one that helped the members define and contrast the domain with other domains – as measured by the team's ability to derive, for the first time, useful high-level assertions.

The lower scores for MAP/A for Implementability and Usability both are based on the need for more specialized training for MAP/A than for AFP. However, this training reflected the background of the Project team in structured analysis. An organization with a staff experienced in object-oriented technology might find MAP/A required less training than AFP.

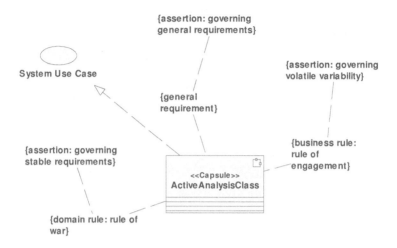

Figure 69: Volatile variability for assertions

Figure 6: Assertions from MAP/A (Redacted)

Future Research Directions

Future research would be most useful in the areas of experiments with the further

application of MAP, studies based on the extensions of MAP, and additional extensions

to MAP (e.g., User-Friendly Access to Formal Methods, Generation of Ontologies and

Taxonomies and Extended Automation of Diagram and Code Generation).

Experiments and Case Studies with MAP

Application of MAP

Application of MAP to additional systems in the AIS domain, to systems in the C2S

domain, Anti-Ballistic Missile domain, and other domains would provide a fuller

demonstration of MAP's feasibility and effectiveness. Doing so in parallel with a

traditional structured analysis process would provide metrics beyond those used in the

evaluation of MAP, which were limited to metrics related to domain rules. Ideally, the

comparison would be done as an experiment, applying alternative processes (the

independent variables to be changed, or factors) to the same system (the object), where

the values of the factors (i.e., specific processes such as MAP) would be the treatments,

with the people (subjects) applying the treatments assigned at random [257]. Multiple

tests (or trials) should then be run for each combination of treatment, subject, and object

[257] to collect metrics and assess the consistency of results across the multiple tests.

Because of the substantial cost of such tests, a more likely future research project would

be a case study of the application of MAP to a complete domain or subdomain.

Studies of the Extensions to MAP

Cognitive studies related to the degree of improved understanding of the system related

to the Meta-Artifact and its properties and qualities such as the Current property and the

Active semantic chain quality would complement the results of experiments and studies

of the additional application of MAP. Studies measuring the effectiveness of the

distinction between needs and requirements, of tacit knowledge, and of knowledge

management principles (see Appendix B, Needs, Tacit Knowledge; Chapter 1, Overview and Problem Discussion sections) in identifying requirements would also complement the other studies and experiments.

Case studies of the cost savings and reliability improvement derived from the Bifurcated Architecture would also complement the results of experiments and studies of the additional application of MAP. Implementing a system with and without a Bifurcated Architecture would allow the comparison of both development and maintenance costs as well a comparison of reliability.

Additional Extensions to MAP

User-Friendly Access to Formal Methods

The formality of the UML specification, with the availability of the Object Constraint Language (OCL), supports semiformal methods. Such support allows formal specifications to be assigned to individual elements of the visual model (as in Figure 70), enhancing the Visual Verification and Validation (see Animated, rather than static, views of development artifacts section). Finally, UML and formal methods intersect to overcome the common barriers to the use of formal methods, as discussed below.

Formal methods have the advantage of proving that a system is correct in the general case, rather than just the particular cases used to test it. Of the reasons cited for why formal methods are not widely used, one that accounts for many of the others is the unfamiliarity of formal methods to most practitioners in systems development. For example, this contributes greatly to the high cost of formal methods [91, 108, 118, 119, 131, and 158]. Many of the notations used for formal methods are complex, with obscure and intuitively ambiguous semantics, especially when the practitioner does not use them regularly [91, 158, and 264].

Unraveling the meaning of a statement in one of these languages for formal methods is sometimes like a complex jigsaw puzzle [146]. This complexity goes back to the origins of formal methods in symbolic logic. Long before computers, Poincare noted, "The difficulty of following a long sequence of analytical steps…" that are "'… deprived by abstraction of all matter [192, p. 121],'" when "'… words are used no longer, but only signs [192, p. 143],'" … which are devoid of the meaning [192, p. 147]."

Most practitioners do not have the time or the inclination to do the unraveling or the raveling. Lack of use tends to make it difficult to standardize and improve the languages of formal methods. The comprehensive tools sets now available for specifying systems with the Unified Modeling Language (UML) offer a means to overcome this difficulty [118, 119, 158, and 264].

The Object Constraint Language (OCL) is part of the current UML specification. The most widely used and comprehensive tools for applying UML also generate up to 85% of the code for a typical distributed, real-time application directly from high-level design and detailed state behavior, that is, from the visual model represented in UML. While these tools identify preconditions and postconditions for operations, they are not included in the automated syntax checking or code generation. Preconditions, postconditions, and invariants can be included throughout the UML model using these tools, but only as documentation (as shown in Figure 70). Full support of OCL with automated syntax checking would add at least semi-formal methods to the code generation.

The next step toward formal methods would be to detect local inconsistencies between the OCL statements and the model. For example, one of the consistency checks should be whether actions to be taken have been specified when a precondition, for example, is not met. The next step should be to determine if the model correctly implemented the specified actions or to note that the model checker could not make such a determination. Finally, these local consistency checks should be expanded to determine if the OCL statements were globally consistent, e.g., one statement might contradict another.

While such checks would fit directly into the current tools, complete implementation of formal methods would require extensions to both OCL and the tools that go beyond their current intent, so supporting formal methods in this way would be a significant research and development effort. To provide user-friendly access to formal methods, tools should stay within the philosophy of OCL, "… to be both formal and simple: its syntax is very straightforward and can be picked up in few minutes by anybody reasonably familiar with modeling or programming concepts [146, Foreword]."

305

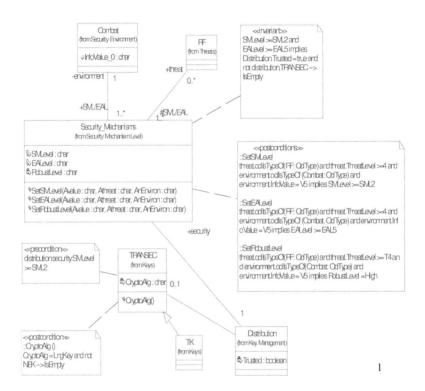

Figure 70: Formal Specifications

Using a simple syntax for formal specifications in conjunction with UML would take
advantage of the broad use of UML, making the formal specifications an integral part of a
widely used standard. This is a key aspect of assuring user-friendly access, because

practitioners who apply UML to a broad range of needs in systems development would find the syntax for formal specifications familiar through their knowledge of UML. Practitioners could write the formal specifications using the standard look and feel of their modeling tools in the semantic context of the model.

Given this ease of use integrated with the model, practitioners could apply formal methods to just those elements of the model where they were most appropriate. This would address perhaps the second biggest reason – along with unfamiliarity – why formal methods are not widely used. Many parts of a system may not warrant the additional effort of formal methods. There are times when the precision of formal methods may be needed to eliminate ambiguity. They force us to say what was assumed [192]. You must state it to see it clearly. But paraphrasing Poincare's question – '... is it always necessary to state it so many times? [192, p. 33]" - must we use it for everything?

Formal methods also have inherent limitations. They are extremely effective for assuring that a system does something right (e.g., by eliminating ambiguity, as noted), but they have no inherent strength in assuring that developers are doing the right thing – eliciting requirements completely and correctly [107 and 91], or dealing with tacit knowledge. The incomprehensibility of formal specifications to all but the experts could make working with domain experts who provide the requirements more difficult [91]. Once again, Poincare anticipated this situation, concluding that the logistic approach is limited and a dead end in many ways, since it allows only for analysis from a closed set of "... undefinable notions and undemonstrable propositions[192, p. 163]." Such an approach,

Poincare thinks, closes off the open-ended potential of intuition. It uses "... a language inaccessible to the uninitiated [192, p. 286]." Yet, in the interest of time, the users are inclined "... to bow before the decisive affirmation of the adepts [192, p. 143."

UML, through use cases and the rich, intuitive semantics of visual modeling with multiple views for different stakeholders is recognized for its strength in capturing requirements. The automated tools amplify this strength by enabling animated visual models (see Animated, rather than static, views of development artifacts section) and evolutionary prototypes (see Kinds of Prototypes section) that are early views of the objective systems rather than disposable prototypes that are not actively linked to a stable architecture or the objective system. Embedding support for formal methods in the UML representation of the system would make its semantic power immediately available for both the raveling and unraveling of the unambiguous but unintuitive formal languages.

These same semantic and dynamic modeling capabilities are also effective in dealing with another limitation of formal methods, which is that they "... cannot yet model nonfunctional properties like performance, reliability, maintainability, ... availability [107, p. 12]" ease of use, structure, and architectural style. Animated, visual models allow stakeholders to determine ease of use and to assess performance (see Animated, rather than static, views of development artifacts section). The visual model directly supports analysis of the structure and style of the system's architecture.

Applying formal methods to parts of a system but not to others [131] would require multiple methodologies (a related approach is referred to as multiformalism in [215]), compounding the complexity related to unfamiliarity. However, if these methods were included with other software creation tools, in an integrated set of tools for the software development environment, without obscuring the semantic power of the overall representation language used by the environment, their value would increase. Or, in Poincare's terms, the developers would still have the ability to seek and easily find the intuitive origin of these formal conclusions, "... letting the soul of the fact escape the machine [192, p.148]."

Combining formal methods with UML modeling, to allow the practical use of formal methods only where needed within a system, would capitalize on the strengths while overcoming the major limitations of formal methods. Developers could then use formal methods for components of the system concerned with life threatening, critical security, or other high-risk issues. Such components often require certification and accreditation, for which formal methods are especially recognized for their effectiveness (e.g., the "Common Criteria for Information Technology Security Evaluation [53]"). Figure 70 shows an example for security management.

Generation of Ontologies and Taxonomies

Tools are available to integrate with the modeling tools used for MAP that will generate ontologies and taxonomies from the classes, relationships, and interactions described by the architecture produced by MAP and embodied in the Meta-Artifact (see Architecture Centricity section and Inference Engine Considerations section). These ontologies and taxonomies could be used to guide the development of multi-agent systems and Semantic Web applications to extend the Meta-Artifact.

Extended Automation of Diagram and Code Generation

While there is strong tool support for managing the use cases with their event flows, linking them to their collaborations, and creating related visual models, the actual translation from informal textual event flows to the formal collaboration represented in UML remains manual in tools such as IBM Rational Rose Technical Developer (RRT). This translation could be readily automated by imposing a formal syntax on the event flows. The value of doing so appears problematic for two reasons.

First, the tools support all but the thought process of translating the event flows into classes and relationships, so there is little or no drudgery involved. Second, the informal textual event flows are semantically rich and remain a vital link in communicating with domain experts. Imposing a formal syntax would at least reduce the familiarity to the domain experts and possibly the semantic content as well. Doing both by having a formal syntax for event flows into which the developers could translate the textual event flows obtained directly from the domain experts sounds like the best of both worlds, but this is in effect what the developers do now when they translate the event flows into their formal UML representations.

RRT currently allows activity diagrams (similar to the traditional logic flow diagram) to be applied to many of the model elements for documentation (e.g., use cases and operations). As documentation, the activity diagrams are important for understanding the complex underlying semantics of the model elements. In the case of operations (i.e., member functions of classes), generating code from the activity diagram for an operation would significantly extend the degree of code generation from visual model elements. The remaining need for manually written code is within states. Generating code from activity diagrams within states would fill in the remaining gap in generating code from the visual model. The improvement in ease of understanding the code within states and operations within classes could easily be more important than the savings in time to manually generate the code, because the states and operations already break the code into small, cohesive chunks and provide the structure into which the code is placed, similar to adding formulas to cells in an existing spreadsheet.

Conclusions and Contributions

MAP with its enabling extensions, and the infrastructure of the enabling technologies on which MAP depends, provides a complete Procedure for developing software intensive systems (see Chapter 3). As a working example of the application of the complete Meta-Artifact Process (see Chapter 3), the prototype demonstrates each property and quality of MAP (see Table 23 and Table 24). The prototype, in using all of the enabling technologies and extensions of MAP (Table 1) demonstrates the feasibility of applying them as hypothesized (see Formal Research Hypothesis and Table 25). The evaluation, at least for Domain Rules Analysis, one of the three extensions of MAP, supports the argument (see Contrasting the Meta-Artifact with Artifacts of Other Processes section) that MAP is an improved process compared to a traditional structured analysis process (AFP, to which MAP was compared, via MAP/A; see Evaluation section). Domain Rules

Analysis is the key extension in MAP for the first capability of the Formal Research Hypothesis. Meeting this capability depends on having a means of establishing boundaries for the domain and staying focused on the problem domain sufficiently to avoid moving prematurely to the solution space (see Table 25). The evaluation (see Evaluation section) showed the value of domain rules in accomplishing this comprehensiveness condition for the Anti-Ballistic Missile domain. The prototype also satisfied this condition by capturing the domain rules (see Table 16) and use cases (see Figure 25) to cover the AIS domain as early artifacts in the prototype's Meta-Artifact (see Chapter 4 and Chapter 5).

Because it builds on mainstream technologies and concepts, MAP is ready for application by a wide range of practitioners, at least in the two domains discussed, but probably others as well. It is especially well suited to green-field projects, because of its domain-wide, problem space focus, which many of the joint command and control projects are. This is an example of one of the fundamental and philosophical characteristics of MAP. MAP is a comprehensive process that uses rigorous analysis (based on current widely accepted methods, concepts, and best practices) to develop an architecture that is appropriate for the domain. It does not impose a predetermined architecture (such as a multi-agent architecture) or pattern on a domain. It does the analysis first to determine what frameworks and patterns are appropriate, which does not preclude using existing ones. Nor is MAP brittle. It is open to changes in the enabling technologies on which it is based. Its unique concepts, such as domain rules and volatile variability, which underlie Domain Rules Analysis and Bifurcated Architecture, are independent of the particulars of the enabling technologies.

MAP can address significant control needs of highly integrated, software intensive systems in network-centric environments (see pp. 61, 100, 146, 158, 161, 178, 220, and 281). The Meta-Artifact offers properties and qualities that help stakeholders and developers understand and reason about domain and target systems of interest. MAP, through the central role of the Meta-Artifact, offers new ways to look at systems and their development, even changing the roles of developers and stakeholders (see Contrasting the Meta-Artifact with Artifacts, Impact of Bifurcated Architecture, and Impact of MAP sections). Domain Rules Analysis, through the Meta-Artifact, can help keep the focus on the problem space long enough to develop a framework and patterns for reuse for two of the use cases in the AIS Domain (see Figure 25 and p. 287). In principle, MAP could be applied in the same way to extend the framework to the other use cases in Figure 25. Given the similarities between control needs for the two domains considered, MAP would likely work well in the C2S domain. Figure 1, by showing the common function of control in the management of enterprises in general, suggests that MAP would apply to any domain involving such management activities.

The separation of stable commonality and variability from volatile variability in the Bifurcated architecture can be achieved through readily available COTS products. The externalization of the volatile variability in the form of business rules in distributed databases allows the volatile variability to be managed by end users (see Chapter 5 and

Appendix C). The combination of the three extensions contributes significantly to reuse, reliability, and interoperability (see Table 27). As Table 27 shows, the three persistent problems are closely related.

MAP causes significant changes in the roles of stakeholders and developers (see Contrasting the Meta-Artifact with Artifacts, Impact of Bifurcated Architecture, and Impact of MAP sections):

- Volatile variability managed by end-users rather than developers
- Domain-wide, long-term focus for development of the first Meta-Artifact and its continued use through MAP
  - o New target systems not a separate standalone activity
  - o Manipulate current Meta-Artifact to produce new target systems
  - o New target systems take account of future target systems
- Responsibility spans phases
- Use entire active semantic chain with multiple views rather than multiple products related to time and discipline

Combining formal methods with visual modeling has the potential to mitigate the major obstacles to their widespread use. Such a combination could lead to making formal methods a standard tool for developers.

MAP introduces the following new concepts, listed under the three extensions of MAP, which are themselves new concepts:

- Meta-Artifact

  o Active semantic chain for systems (see Appendix B)

  o Special views as byproducts (see p. 93)

  o Process and product as different views (see p. 96)

  o Problem space and solution space as different views (see p. 122)

  o Visual verification and validations (see p. 91)

  o Treating states on state diagrams as finer grained encapsulation within the encapsulation provided by the active class for reliability and collaborative development (see p. 220)

- Domain Rules Analysis

  o Domain rules (see Domain Rules section)

  o Volatile vs. stable variability (see Domain Rules section)

  o Uses concept of volatile variability to relate business rules to commonality and variability, making the related analysis techniques available to the analysis of business rules (see Appendix B, Commonality and Variability)

  o Uses dichotomy between commonality and stable variability versus volatile variability throughout the system's development and implementation to identify volatile variability (see Domain Rules and Procedure sections)

- Bifurcated architecture

  o Physically separates volatile variability at points of logical separation throughout a domain, without imposing a rule-based or business-rule-centric architecture (see Bifurcated Architecture to Encompass

Commonality and Variability section)

- o Relates controls and business rules by defining controls as the prescriptive

  rules to enforce declarative business rules (see Table 20 and Table 21)

- Needs versus requirements

  - o Distinguishes tacit requirements from explicit requirements (see Appendix

    B, Needs and Tacit Knowledge)

  - o Applies knowledge management theory to uncovering tacit requirements

    (see Chapter 1, Overview and Problem Discussion sections)

- Representation of the standard management process as a first-order cybernetic

  feedback (closed-loop) system (see Figure 1 and Chapter 1, Overview section),

  allowing the application of systems theory to the control component of the

  standard management process. In turn, this representation highlights the

  commonality of controls in multiple domains involving diverse organizations,

  e.g., business and military enterprises in the AIS and C2S domains.

Appendix A: MAP and Persistent Issues in Software Development

Table 27: Enhanced Reliability, Reuse, and Interoperability

	Reliability (1)	Reuse (2)	Interoperability (3)
1	Semantic chain of the Meta-Artifact improves understanding for development, operation, and maintenance (pp. 156, 119, 170, and 187], leading to fewer defects that result from inadequate knowledge of what the system is supposed to do	MAP enables derivation of stable frameworks and patterns to generate components from which a wide range of target applications can be composed (see Definition of Domain Rules Analysis section), increasing reuse of frameworks, patterns, and components	Externalization of volatile variability (see cell 4(2)) increases the range of applications for which a component would be a candidate to interoperate with by increasing commonality (see cell 4(2))
2	Externalization of volatile variability through the Bifurcated Architecture (see cell 4(2))reduces defects by requiring fewer changes to existing code (pp. 155 and 282)	Domain Rules Analysis enables creation of Meta-Artifact that represents totality of solution space (see Comparison of Other Domain Analysis and Visual Modeling Processes to MAP, Figure 14, and Role of the Meta-Artifact in MAP), from which to derive stable frameworks and patterns to generate components from which a wide range of target applications can be composed, increasing reuse of frameworks, patterns, and components	Composition of a wide range of target applications (see cells 1(2) and 2(2)) from the same frameworks and patterns increases the number of components with inherent interoperability
3	Architecture Centricity (p. 176) improves reliability by increasing reuse (see cell 3(2)), allowing proven frameworks, patterns, and components to be used rather than new or modified ones (see pp. 130 and 256 for composing with pub and sub interfaces)	Architecture Centricity promotes reuse (p. 176) by emphasizing components and how they interact, especially when they interact through standard interfaces (pp. 130 and 256 for composing with pub sub interfaces)	
4	Domain Rules Analysis improves reliability by extending reuse (see	The Bifurcated Architecture externalizes the volatile	

	Reliability (1)	Reuse (2)	Interoperability (3)
	cell 2(2)), allowing proven frameworks and patterns to be used rather than new ones, or proven components to be used rather than new or modified ones (p. 129)	variability captured in business rules, increasing the commonality of components to enable wider reuse (see Domain Rules section)	
5	Externalization (see cell 4(2)) and distribution of business rules among multiple repositories reduces risk of unauthorized changes and cascading errors (p. 280)		

Table 27 shows the contributions of MAP to reliability, reuse, and interoperability

beyond those provided by the existing body of knowledge, represented by the enabling

technologies. It contains the enhancements to reliability, reuse, and interoperability

provided by the three extensions to MAP, but not those provided by the enablers (see

Table 6), such as inheritance, encapsulation, and polymorphism. Table 27 highlights the

interdependence of reliability, reuse, and interoperability, e.g., see the cross references in

a number of the table's cells. The larger number of rows in the Reliability column in

Table 27 also indicates that MAP is most strongly related to reliability, a key component

of controls

Appendix B: Glossary

All of the entries in this glossary define only how the terms are used in *Breakdowns*, so there may be other definitions for these terms not included here. The symbol † indicates a new concept developed in *Breakdowns*.

Active Semantic Chain (†): consists of the highest-level semantics of the system – requirements articulated as artifacts (see Background of the General Problem section above)– electronically connected through successive links in the chain – e.g., analysis and design artifacts – to the lowest-level semantics of the executable components (e.g., software code) and their outputs. It is created by the integrated tools from the electronically stored artifacts, building on the semantics of the common representation provided by UML. Embodied in the Meta-Artifact, the active semantic chain preserves the semantics of the system, from its genesis to its retirement.

Using the Meta-Artifact, the active semantic chain provides bi-directional narratives – both to articulate what stakeholders know explicitly and to trace back to the rationale for what they know implicitly (experience-related knowledge or common sense that are tacit, or background, knowledge).

The active semantic chain is distinctly different from the relationship among artifacts found in practice. As discussed in Contrasting the Meta-Artifact with Artifacts of Other

Processes, in practice, requirements, analysis, design, implementation, and operations are physically and conceptually separate activities.

The active semantic chain is at the core of the Meta-Artifact's ability to promote understanding of the system. A key aspect of this understanding is documentation that is always current, temporally and spatially accessible, and represented from the stakeholders' viewpoint. Such documentation has been the dream and the nightmare of systems development since the first question about how a system worked. Because the integrated tools provide views tailored to the interests of stakeholders and based on all of the artifacts from which the current running system is derived, the documentation is far more comprehensive than what is available in current practice.

The importance of this semantic chain and its many uses are described throughout *Breakdowns*, beginning with its foundations in the philosophy of science. Philosophers of science have written about the importance of capturing the genesis of ideas and the semantic chain of the artifacts that trace a development, whether a mathematical proof or a cultural tradition, from inception through to fruition. E.g., William Dilthey [69] concludes:

> The study of the concepts and precepts of existing cultural systems suffers from their not having the process that led to them '... preserved in its original fluid form' but rather 'objectified and compressed in the smallest possible form', i.e., in the shape of legal concepts.

Or as mathematician Henri Poincare observed in [191], we need to see "...the genesis [often a flash of intuitive insight] of our conceptions, in the proper sequence."

Adaptability: support for change within current the domain, e.g., for new target systems, changes to existing target systems, or technology insertion. These would be monotonic changes, based on the original domain rules analysis.

Background: a term used in Patriotta's knowledge management framework to refer to the tacit, unstated, and taken for granted assumptions underlying both individual and social practices [188]. With the passage of time, tacit knowledge from different periods forms layers in the background, with less likelihood that stakeholders will recall the explicit knowledge behind the tacit routines, or that a narrative will exist from which to recover the explicit knowledge (see **Information, Explicit Knowledge, Narratives,** and **Time** below).

Bifurcated Architecture (†): separates the **volatile variability**, normally embedded in code, into an external database so that authorized users can directly change it. Volatile variability, also referred to as **Business Rules**, can be maintained more efficiently and accurately in an external database than when it is embedded in code and is more accessible for such purposes as analysis, audit, and training. The Bifurcated Architecture element of the Meta-Artifact Process (MAP) is orthogonal to the architecture-centric element of MAP because it is applied to the architecture derived through MAP, rather than driving the architecture development. This orthogonality preserves the architecture-centricity of MAP based on all requirements, in contrast to other approaches to handling rules, e.g., rule-based or rules-centric approaches that drive the architecture development.

Bifurcated Architecture extends the domain analysis concepts of commonality and variability in two ways:

- Differentiating stable variability from volatile variability
- Providing physical as well as logical separation

Breakdowns: one of three lenses, along with **Time** and **Narratives,** of a phenomenological approach to the study of knowing in the context of organizing (see also **Information** and **Explicit Knowledge**). The three lenses provide operational devices to reveal the tacit, unstated, and taken for granted assumptions underlying organizational practices against which organizational knowledge (a form of passive knowledge) is utilized by the organization's stakeholders on a day-to-day basis. Explicit Knowledge that is in the **Foreground**, e.g., when a system is under development (the social phenomenon of the development of the system is conspicuous as it is witnessed for the first time under the special circumstances of the system's conceptualization, design, and implementation) becomes part of the background (used by habit – the knowledge is deeply internalized and institutionalized so that it is used in an almost automatic and irreflexive way – or unused) after the system is in operation and performs tasks for stakeholders so that the stakeholders' knowledge of how the system performs those tasks is no longer conspicuous.

Breakdowns disrupt the system's routine operation (discontinuity in action), forcing stakeholders to again recall how the system performs its tasks. In this way, breakdowns provide a lens to focus on what would otherwise be tacit knowledge [188 and 195], e.g., embedded in the system and taken for granted. This focus discloses intentionality (why) and highlights the cognitive dimension (empirical how) of knowledge embedded in the system [188]. The Meta-Artifact, as a narrative articulated as artifacts (text) of the knowledge creating dynamics of the system's development (see **Narratives**), through the **Active semantic chain**, provides a means to convert information back to the knowledge needed to repair breakdowns (see **Time**).

Business Rules: represent the volatile variability of the application because they change in response to such things as the current situation – e.g., environmental conditions, technology, knowledge, and attitudes. Unlike domain rules, a system's business rules are under the control of the organization that uses the system. They capture the organization's business philosophy and practices in terms "... that describe, constrain, and control the structure, operations, and strategy" of the organization [126].

Business rules may be viewed as a constraint on particular applications in a domain. They may also be viewed as the decision-making rules for the application, within the invariants of the domain (represented by the domain rules). They are the same rules needed by decision support tools.

Business rules may be derived from external sources that the organization does not control, such as domain rules, regulations, and cultural considerations. This type of business rule reflects the organization's interpretation of how to comply with such external sources. The organization's interpretation may change both in terms of how to comply and of which external sources are relevant. To the extent that the external source is itself subject to change (e.g., frequently revised federal regulations), the volatility is increased. Business rules may change very frequently (e.g., hourly or daily for online sales) or only a few times over the life of a system. In contrast, domain rules are unlikely to change over the entire life of the system. Regardless of the exact frequency, when business rules do change, externalizing them avoids costly maintenance activities.

Command and Control: the exercise of authority and direction by a properly designated commander over assigned and attached forces in the accomplishment of the mission. Command and control functions are performed through an arrangement of personnel, equipment, communications, facilities, and procedures employed by a commander in planning, directing, coordinating, and controlling forces and operations in the accomplishment of the mission [65, glossary].

Command, Control, Communications, and Computer Systems: integrated systems of doctrine, procedures, organizational structures, personnel, equipment, facilities, and communications designed to support a commander's exercise of command and control across the range of military operations [65, glossary].

Commonality: similar functions and attributes across a group of applications. In OOT, these similarities are captured in derived classes through inheritance from base classes and through aggregation. Contrast with **Variability** and see **Domain Rules**.

Control: verifying whether everything occurs in conformity with the plan adopted, the instructions issued, and the principles established, and then taking the appropriated corrective actions [76]. Many of the control needs of a system are captured in business rules.

Controls (†): by their nature, controls must test facts to determine if prescribed results have been met. That is, controls must enforce the constraints of the system. The controls considered by *Breakdowns* are the prescriptive rules implementing the business rules.

During analysis and design, business rules would be identified for each active class. They would be converted during implementation to production rules to be executed by an inference engine, rather than program code. The production rules would be the mechanism for implementing the prescriptive instructions of the controls that enforce the business rules; they would be maintained in external files assigned to the components whose business rules they would enforce. The separation of the rules for each component would support a fundamental control procedure, separation of duties. Users authorized to maintain rules in one database could be physically prevented from changing rules in another database. Cohesion and separation can be further improved within the same database by using multiple packages for rules.

Control Background: that part of the **Background** in which tacit control needs are embedded, including systems that intentionally embed explicit knowledge for the purpose of carrying out tasks otherwise performed by people, accelerating the conversion of explicit knowledge to tacit knowledge by adding to the conversion caused by routinization of human tasks [188].

Domain Rules (†): the invariant rules that apply to systems in a domain, based on the underlying principles, theory, or traditions of the domain. Examples are doctrine in the military, information theory for engineering, and duality (debits and credits) for accounting. All applications (systems) for the domain must take account of the domain rules to determine which apply. Not all domain rules apply to every application of a domain, but each application must incorporate at least one of the domain rules in order to belong to the domain. The differences in domain rules applicable to a system reveal a partitioning of the domain.

Domain rules form the overarching category of requirements that characterize a domain by distinguishing it (and the systems that meet the needs for the domain) from other domains. Domain rules serve as meta-rules that govern what subordinate categories of requirements are appropriate for the domain. In this way, domain rules are constraints. Domain rules set the boundaries of a domain in pure problem space terms (see Definition of Domain Rules Analysis section), by helping to identify requirements that lie inside the domain and that must be captured in systems for the domain.

On the one hand, domain rules are constraints. Domain rules help set the boundaries of a domain in pure problem space terms. On the other hand, domain rules are part of the commonality (see **Commonality**) of the domain. The commonality drives reuse, especially with OOT, where commonality can be incorporated through inheritance.

Domain Rules Analysis (†): uses **domain rules** to help set the boundaries of the domain and, in conjunction with the volatile variability captured for the Bifurcated Architecture, to keep the focus on the problem domain until a comprehensive analysis is completed.

Explicit Knowledge: active knowledge, embedded in the human consciousness, in contrast to passive knowledge that is written down, printed on paper, or stored on electronic devices and referred to as information [170 and 266].

Extensibility: Supports domain evolution, e.g., through accretion of other domains, environmental change, boundary expansion through new technology, or mandated boundary expansion through social, economic, or political change. This evolution by definition would be non-monotonic, in that the new aspects of the domain did not exist when the original domain was analyzed and would necessarily involve new concepts [7].

Foreground: a term used in Patriotta's knowledge management framework to refer to explicit knowledge in contrast with **Background**, which refers to implicit or tacit knowledge that is experience-related, applied unconsciously, or taken for granted [188].

Framework: the set of class hierarchies for the domain and classes aggregated with classes in the hierarchy. Customary definitions using object-orientation include patterns as part of the framework, but to reduce ambiguity, *Breakdowns* considers frameworks and patterns separately. The generalization-specialization relationship of hierarchies establishes the commonality (superclass) and variability (subclass or class from another hierarchy) among classes, which is a focus of OODA. Aggregated classes are included because, like generalization-specialization, aggregation only elaborates the definition of a class, unlike associations or the various dependency relationships that are used for patterns.

Information: passive knowledge, written down, printed on paper, or stored on electronic devices, in contrast to the active knowledge embedded in human consciousness [170].

Intelligence:

- The product resulting from the collection, processing, integration, analysis, evaluation, and interpretation of available information concerning foreign countries or areas

- Information and knowledge about an adversary obtained through observation, investigation, analysis, or understanding [65, glossary].

Meta-Artifact (†): the electronically linked set of all of the artifacts of development, amplified by three extensions and five enabling technologies and its recursive use in the Meta-Artifact Process (MAP) for its own development. The Meta-Artifact provides a knowledge management **Narrative** about a system that makes the background explicit. Narratives in knowledge management serve as a basic organizing principle of human cognition [188]. The Meta-Artifact continuously supplies the teleological remedy, through the active semantic chain, to the entropy that otherwise occurs with time, as development progresses and during operation and maintenance of the system, through its application in the Meta-Artifact Process (MAP), thereby converting knowledge captured in the Meta-Artifact into action and action into additional knowledge in the Meta-Artifact. The Meta-Artifact thus preserves the knowledge of the system and embedded in the system as explicit knowledge, preventing the knowledge from receding into history (becoming forgotten with the passage of time) and becoming tacit. By providing a means of identifying tacit knowledge contained in implicit domain rules (see Procedure section), MAP helps convert previously tacit knowledge into explicit knowledge. The Meta-Artifact preserves the ontology of the problem and solution spaces.

The Meta-Artifact extends the modeling concept that the model is the application [160] by applying the concept to entire domains. The Meta-Artifact also extends the particulars of the concept beyond that of automatically generating code from the visual model (the basis for saying that the model is the application). That is, automatic code generation is just one of the sub-qualities noted for the Meta-Artifact (see Animated, rather than static, views of development artifacts section).

More generally, the Meta-Artifact contrasts with the collection of artifacts produced in practice by methodologies other than MAP, including those produced for visual models from which code is generated, because of the interdependence it has with Domain Rules Analysis, Bifurcated Architecture, and the five enablers. In particular, discussions of visual models in other methodologies do not refer to an active semantic chain with a central role in a development process designed to comprehensively analyze a complete domain (problem space) independently of the solution space, based on domain rules (see Domain Rules Analysis section).

The Meta-Artifact provides the rules and structure for its own continued development (see Role of the Meta-Artifact in MAP section). That is, the Meta-Artifact is used recursively in the incremental building of itself by MAP, analogous to the metacircular role of the UML metamodel [20], except that the UML metamodel is the starting point of UML, whereas the Meta-Artifact is the goal of MAP. The Meta-Artifact goes beyond the role of meta-models and meta-data, which provide the formalisms for defining other models or data, because it includes the completed artifacts themselves. MAP converts knowledge captured in the Meta-Artifact into action and action into additional knowledge in the Meta-Artifact.

Narratives: one of three lenses, along with **Time** and **Breakdowns,** of a phenomenological approach to the study of knowing in the context of organizing (see also **Information** and **Explicit Knowledge**). The three lenses provide operational devices to reveal the tacit, unstated, and taken for granted assumptions underlying organizational practices against which organizational knowledge (passive knowledge) is utilized by the organization's **Stakeholders** on a day-to-day basis.

> Narratives, articulated as texts, can be seen as material traces of learning and collective remembering processes, social imprints of a meaningful course of events, documents and records of human action. They allow people to articulate knowledge through discourse [188].

Stakeholders' knowledge of control needs (requirements) is articulated through narratives and the narratives articulated as artifacts (the system development equivalent of articulating narratives as text in knowledge management) for development and maintenance of the systems. For brevity these multiple steps for the transformation of control needs, whether explicit or tacit, into system artifacts (including code) are referred to collectively as articulating control needs (or knowledge) as artifacts.

When articulated as text (artifacts in the case of system development), narratives allow the tracing of the process of articulating stakeholders' knowledge, including experience-related (tacit) knowledge, into the system through development artifacts. The narrative brings out the experience-related knowledge as part of the discourse conducted among stakeholders by articulating it along with other knowledge elicited from them. If the experience-related knowledge is challenged, e.g., by other stakeholders, additional narratives can be used to reconstruct the rationale underlying the experience-related knowledge, revealing the explicit knowledge that had become habit (see **Breakdown**) or whose original purpose had been forgotten as it receded to the background (see **Time**). Breakdowns expose a background that is otherwise taken for granted, e.g., the functions of a system that no longer performs a task or a manual control that is no longer timely, exposing the assumption that the control will handle the event soon enough to avoid unacceptable loss. The layers (see **Background** above) created over time as explicit knowledge recedes to history offer the potential to uncover explicit knowledge in an earlier layer that was the origin of what is now tacit. The narrative lens works in conjunction with the lenses of breakdowns and time in rediscovering explicit knowledge by reconstructing evidence exposed by breakdowns or contained in earlier layers, reversing the knowledge creating dynamics [188].

This two-way application of narratives – both to articulate what stakeholders know explicitly and to trace back to the rationale for what they know implicitly (experience-related knowledge or common sense that are tacit, or background, knowledge) suggests that retaining the narratives (articulating them as artifacts) used in developing systems is worthwhile. During development of the system, the artifacts of development must be accessible and understandable (the social phenomenon of the development of the system is conspicuous as it is witnessed for the first time under the special circumstances of the system's conceptualization, design, and implementation) to be useful [188].

However, narratives or portions of narratives may not have been retained (articulated as artifacts). That is, the narratives or portions of them may have consisted of entirely oral communications, possibly augmented with temporary text, graphics, and demonstrations. The artifacts that are articulated that emerged from preceding narratives (or portions of them) that were not retained become disconnected from their experience-related or explicit knowledge (discontinuity in knowledge), i.e., there are links missing in the semantic chain of understanding for the artifacts. To have a complete semantic chain of understanding, all of the narratives ultimately leading to artifacts must be retained and be accessible to all stakeholders during the full lifecycle of the system. That is, if a narrative (or a portion of it) is retained, but put aside (e.g., stored in a medium or location not accessible to stakeholders) once the artifact it leads to is complete, the semantic chain of understanding is broken, producing gaps in understanding. The Meta-Artifact provides such a complete semantic chain, accessible through its **Active semantic chain**.

Needs: Needs comprise everything the system must do. Requirements are those needs that have been explicitly identified. Needs not explicitly identified as requirements are tacit requirements (see General Problem). For example, there are control needs that must be met by the system of control in order for it to be adequate, even if such control needs are not identified as requirements (see Chapter 1, Overview). Tacit requirements are embedded in tacit knowledge (see **Tacit Knowledge**). The contrast between needs and requirements is important because assuming all needs are contained in the explicit requirements provided by stakeholders and in documents available to developers, such as an external solicitation or internal request (see the section Survey and Review of the Literature of Prior Research below), leads to breakdowns (see **Breakdowns** and Background of the General Problem section).

Pattern: a collaboration as defined in [22].

Problem Space: Consists of the requirements and related information that describe the problem or need to be solved by the system.

Reconnaissance: a mission undertaken to obtain, by visual observation or other detection methods, information about the activities and resources of an enemy or potential enemy, or to secure data concerning the meteorological, hydrographic, or geographic characteristics of a particular area [65, glossary].

Routine: organization as a clockwork, based on successful responses to problematic situations. Once a specific routine has been invented, the problem addressed by it simply stops being a (conscious) problem, so routines become carriers of tacit knowledge [188].

Rule-Based Process: a process that is represented entirely by rules, with the reasoning strategy as well as the rule sets underlying the reasoning made explicit [99].

Rule-Constrained Process: an existing business process, implemented in conventional software code, which needs to have certain constraints enforced. These constraints may be externalized as rules [99].

Scalability: the relative cost, time, and risk entailed for an application to process increased volumes beyond the capacity of the current platform configuration.

Six Sigma: a knowledge-based, comprehensive and flexible system for achieving, sustaining, and maximizing process, service, and product improvement. It incorporates many of the principles and tools found in Total Quality Management and Business Process Reengineering [184]. Six sigma complements Integrated Product Development Systems and Capability Maturity Model guidelines by providing a means to implement new or improved processes, services, and products.

Solution Space: Consists of the system or systems that meet the requirements in the Problem Space defined by the domain.

Stable Variability (†): some variability, such as between cycles in the AIS domain, would be stable because it would be based on the domain rules and other stable aspects of the domain such as traditional functions and processes. In OOT, stable variability would be captured through specialization during design in creating the framework (see Domain Rules Analysis section and Appendix B, Framework), selecting from among the specialized classes in the framework during composition [93 and 95], and activation of services at runtime. The classes in the framework, including the derived classes specialized to capture the variability (Procedure, step 15) for different applications within the domain, would be stable for the same reasons as the program code (see Modeling Considerations in Applying the Procedure section). That is, the variability that distinguishes one application from another is not the source of volatility.

For example, the component to match depreciation expense with sales for a period would not be the same as the component to match direct materials costs with sales, but the portions of each component related to ensuring that appropriate components were invoked to match sales and expense for the accounting period (i.e., the domain rule for *Matching*) would be inherited. Furthermore, the specialized aspects of the two would seldom require change. There are basic depreciation methods that seldom change. What changes is which method to apply to a class of assets, which depends on the business rules.

Stakeholder View: how a stakeholder looks at the system (IEEE std 1471), in terms of what it actually does or should do.

Stakeholder: people who have a stake in the system, e.g., various developer disciplines, end-users, domain experts, managers, regulators, auditors, and certifiers (IEEE std 1471).

Surveillance: the systematic observation of aerospace, surface or subsurface areas, places, persons, or things, by visual, aural, electronic, photographic, or other means [65, glossary].

Tacit Knowledge: implicit knowledge that is experience-related, applied unconsciously, or taken for granted (see **Background** and **Foreground**) and embedded in human routines, organizational culture, and existing systems [188 and195].

Time: one of three lenses, along with **Breakdowns** and **Narratives,** of a phenomenological approach to the study of knowing in the context of organizing (see also **Information** and **Explicit Knowledge**). The three lenses provide operational devices to reveal the tacit, unstated, and taken for granted assumptions underlying organizational practices against which organizational knowledge is utilized by the organization's **Stakeholders** on a day-to-day basis. Explicit knowledge at one time, e.g., when a system is under development (the social phenomenon of the development of the system is conspicuous as it is witnessed for the first time under the special circumstances of the system's conceptualization, design, and implementation), recedes to the background (becomes history and forgotten with the passage of time) at a later time, when the system is in operation and the knowledge utilized by stakeholders to develop the system is no longer conspicuous.

Time provides a lens to understand the dynamics of stakeholder interaction (social becoming) underlying the process of knowledge construction in systems development organizations, i.e., viewing systems development as a social process to convert knowledge into system). Time also provides a lens to see how knowledge entropically becomes information (discontinuity in time [188]). The information may be immediately useful, e.g., system documentation, or require extensive transformation to be useful, e.g. only the external behavior of the system as it is observed or inferred from previous outputs of the system. The combination of understanding the original teleological conversion of knowledge into the system (articulating requirements as artifacts) and the entropic conversion of that knowledge into information provides insight into countering the entropic conversion with the Meta-Artifact process, reverse engineering information to construct a Meta-Artifact after a system is developed, and to use the Meta-Artifact to repair a system Breakdown [188].

Variability: different functions and attributes across a group of applications. In OOT, these differences are captured in derived classes through specialization achieved by overriding or adding to functions and attributes inherited from base classes, by adding new base classes, and through aggregation (Contrast with **Commonality**). Some variability may be stable (see **Stable Variability**), but some may be volatile (see **Volatile Variability**). In the Bifurcated Architecture, volatile variability is allocated to an external repository (see Bifurcated Architecture to Encompass Commonality and Variability section).

Visible: something stakeholders of a system can actually see, such as the output of an automated system or the documents used in a manual system; also, the representations of an automated system that the stakeholders can see, especially a graphical representation or visual model.

Visual Model: a graphical representation of an automated system.

Volatile Variability (†): represented by business rules, volatile variability is in contrast to the stability and commonality of domain rules and stable variability. The concept of volatile variability supplements the customary definitions of business rules by placing them in the context of commonality and variability and aiding in their identification during all phases of the system's lifecycle. That is, developers and stakeholders can apply the criterion of volatility – frequency of change -- to identify functions to be externalized. The volatility criterion should make it unnecessary to have detailed definitions and rules for identifying business rules. While ultimately a matter of judgment as to whether the variability was volatile, the likely frequency of change would provide an objective measure. The frequency might vary from hourly to a few times over the lifecycle of the system, but regardless of the exact frequency, when business rules do change, allocating them to an external repository (see Bifurcated Architecture to Encompass Commonality and Variability section) avoids costly maintenance activities.

Appendix C: Creating an External Business Rules Database

The graphics in this appendix are from ILOG Rule Builder, a product of ILOG S.A.

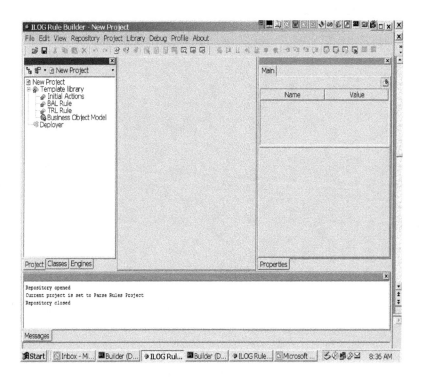

Figure 71: Create a New Project for the Database

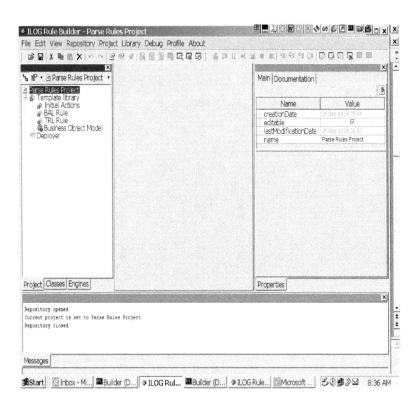

Figure 72: Project Created

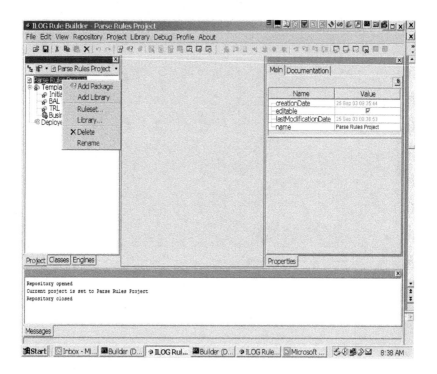

Figure 73: Create a Package for the Database

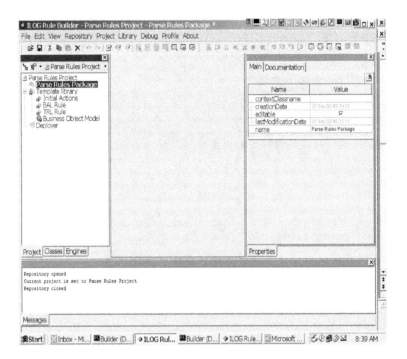

Figure 74: Package Created

(See p. 281 for a discussion of using multiple packages.)

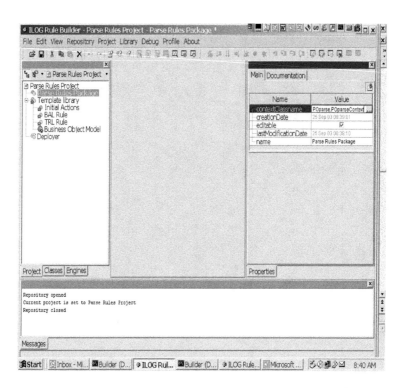

Figure 75: Specify the Context Based on Exported Members

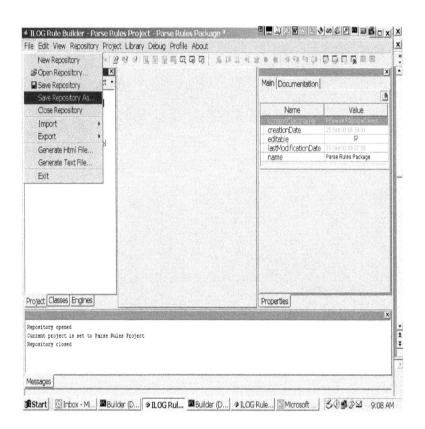

Figure 76: Create the Database

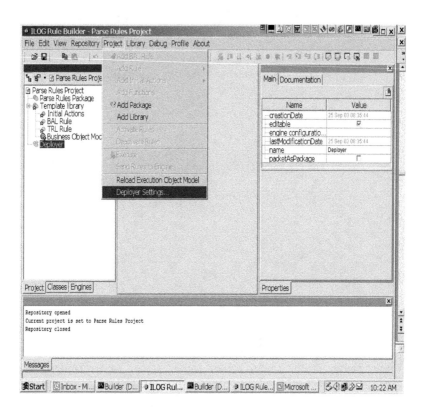

Figure 77: Import a File with Member Variables and Functions

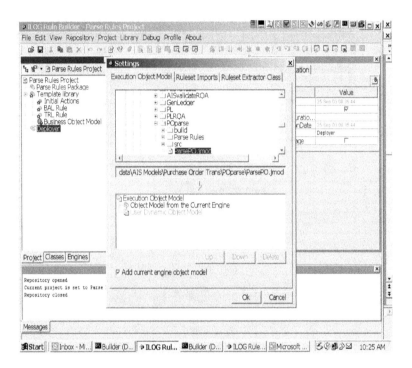

Figure 78: Select File with Member Variables and Functions to Import

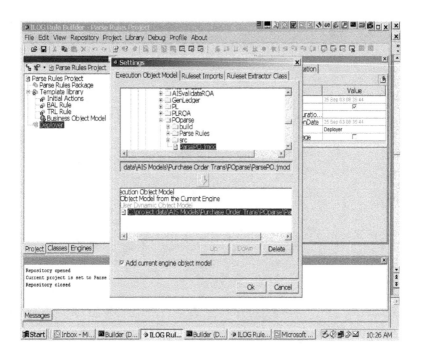

Figure 79: File with Members Variables and Functions Imported

Figure 80: Add a New Class to the Database from Imported File

348

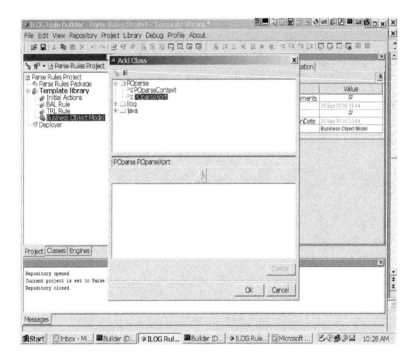

Figure 81: Select the Class to be Added to the Database

349

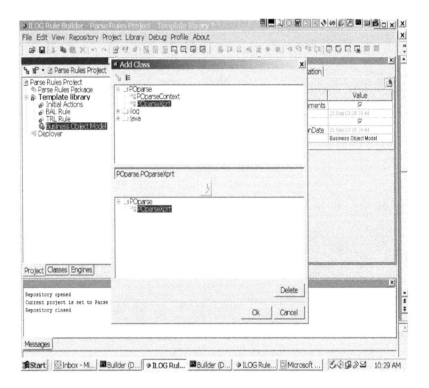

Figure 82: New Class Added to the Database

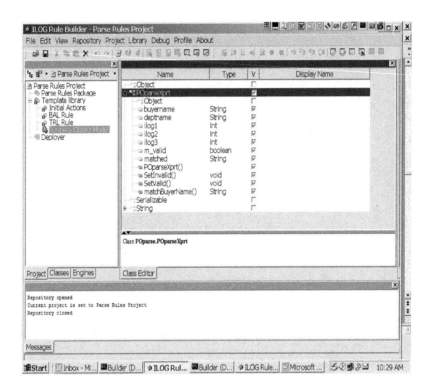

Figure 83: Edit New Class to Add Natural Language Descriptions

351

Figure 84: Natural Language Descriptions Added -- POparse

Figure 85: Natural Language Descriptions Added – POvalidate

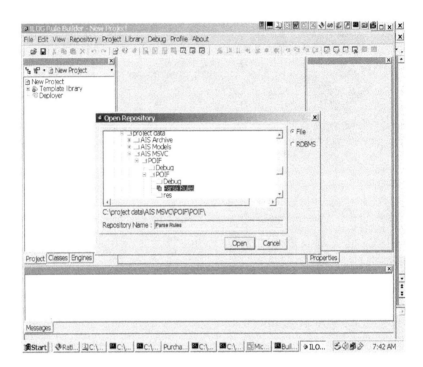

Figure 86: Open an Existing Database – POparse

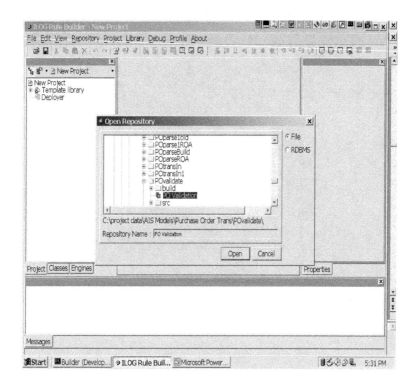

Figure 87: Open an Existing Database – POvalidate

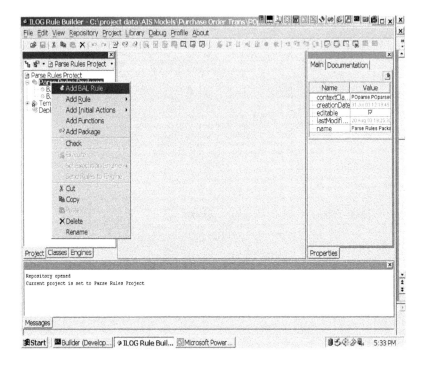

Figure 88: Create a New Rule Template – POparse

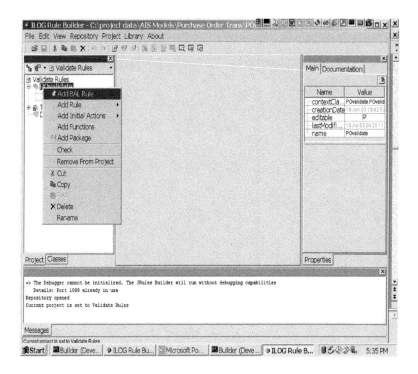

Figure 89: Create a New Rule Template – POvalidate

Figure 90: New Rule Template Created – POparse

Appendix D: Inheritance for AIS Controls

Traditional internal control categories are thoroughly described in the literature, e.g., the Institute of Internal Auditors [205], has the following categories:

- Activity Type
    - ◊ Accounting
    - ◊ Administrative
- Control Objective Type
    - ◊ Compliance
        - ⇒ Financial
        - ⇒ Operational
        - ⇒ Regulatory
        - ⇒ Fraud Prevention [205, p. 112]
    - ◊ Safeguard assets
    - ◊ Reliability and integrity of information
    - ◊ Economy and efficiency of operations
    - ◊ Accomplish organizational objectives
- Technique type [205, p.105]
    - ◊ Preventive
    - ◊ Detective

◊ Corrective

- Method type [205, p. 106]

  ◊ Organizational

  ◊ Operational

  ◊ Personnel

- Control Point

  ◊ Authorization

  ◊ Entry

  ◊ Inventory

Data processing categories would include the following:

- Transaction type

- Management reporting

- OLAP

The above list shows eight major categories on the same level, to emphasize that there is no inherent hierarchy. The existence of such a widely accepted and proven list of categories is an important aspect of the AIS domain for purposes of validating that all requirements of the problem domain have been covered. That is, having at least an abstract class for every category of internal control, which represents the volatile portion of the AIS domain, provides a reliable checklist.

The appropriate hierarchy would require analyzing requirements in each situation, e.g., for a particular customer. Even for a given customer, the appropriate hierarchy might change over time. There is clearly a design tradeoff between organizing the controls as subordinated types, which may be closer to the user's cognitive model, and a flatter structure, which would simplify coding and maintenance.

Figure 91 shows an example of a class structure that implements some of the above categories using a deep hierarchy. The base class for the structure is General Control Objectives [205]. The next level in the hierarchy consists of a class for each category of control objectives. Beneath each category class, two classes distinguish between Administrative (where the focus is on operations without any direct link to accounting records [89, p. 194; 205, p. 98]) and Accounting (where there is a direct affect on the accounting records) controls. The leaf level specifies the particular controls to be implemented, depending on whether they are Financial, Regulatory, Fraud Prevention, or Operational.

The hierarchy in Figure 91 would apply to each transaction type in Figure 67, that is, Purchase Order (PO), Cash Disbursement (CD), and Receive Order (RO). Generally, the type of controls in Figure 91 would only apply to validation, rather than parsing. The parsing rules would be limited to determining whether the transaction was well formed, rather than the content of the transaction. Within Transaction Type, appropriate leaf-level controls would be selected. The prescriptive instructions for the internal controls would be implemented at this leaf-level. Table 28 has an example of prescriptive instructions for each of the four leaf level subclasses.

Although not shown in Figure 91, additional specialization could be applied under each of the leaf-level subclasses to create subclasses for Techniques, Methods, and Control Points. This deep specialization could also be applicable under Administrative Activity as well as under Other Objectives. However, the number of levels in the class hierarchy in Figure 91 is already high, so making it flatter or at least no deeper becomes a consideration for coding and maintenance. One way to do this would be through composition. That is, there would be a class structure for Accounting Compliance Objectives, composed of a class for preventive financial compliance methods for operational concerns for authorization control points, a class for detective financial compliance methods for operational concerns for authorization control points, etc.

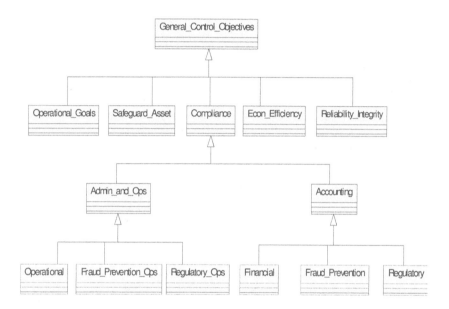

Figure 91: AIS Internal Control Hierarchy

Table 28: Prescriptive Instructions for Internal Controls gives a representative

prescriptive rule for each of the leaf-level subclasses in Figure 91. In practice, there

could be hundreds or thousands of such rules applied by software.

Table 28: Prescriptive Instructions for Internal Controls

Compliance Control Objective Subclass	Prescriptive Instructions
Financial – Compliance with budget policies	If upon approval of this request for purchase order, total encumbered dollars for this subsidiary ledger account would be greater than the budget for that account, reject the request
Operational – Compliance with authorization polices	If the amount for this purchase order exceeds the signature authority of the Buyer (purchasing agent), reject the purchase order
Regulatory – Compliance with tax law and regulation	if the type of asset-type specified for this subsidiary ledger account does not match the asset type for this depreciation method, reject the transaction: either the wrong subsidiary ledger account is being used to set up this asset or the wrong depreciation method has been specified
Fraud – Compliance with legal and policy requirements	Select all transactions for a specified subsidiary ledger account for a specified time period exceeding a specified dollar amount, then process the details of those transactions (e.g., name of vendor, name of purchasing agent, address of vendor, shipping address) through specified neural network to detect patterns of fraudulent activity.

A combination of composition and specialization should also be considered in the example in Figure 91. That is, the Financial Objectives subclass could be composed of classes for Techniques, Methods, and Control Points, or there could be multiple Financial Objective subclasses, one composed with a Techniques class, one with a Methods class, and one with a Control Point class. These are just some of the many possible arrangements, so an extensive analysis effort will probably be required in some cases. The structure in Figure 91 could also be changed. The structure in Figure 91 has the

Administrative and Accounting activity derived from the objectives. It might be just as appropriate in some circumstances to reverse this order.

Of course, whenever the customer has an established hierarchy – explicit or implicit – the analysis would be greatly simplified. Perhaps a research objective should be to capture metrics, to decide first between compatibility with the users' existing cognitive models [254] and ease of coding, then among the various structures that could be used once the first decision was made, and to develop heuristics based on the metrics. Heuristics can be found in the literature, e.g. [89, p. 194] considers safeguarding of assets and reliable information (defined as accurate accounting information in [89]) to be applicable only to accounting controls.

References

[1] Adelman, Leonard. *Evaluating Decision Support and Expert Systems.* New York: John Wiley & Sons, Inc., 1992.

[2] Aikins, Jan. "Business process reengineering: Where do knowledge-based systems fit?" *Expert*, n.d., 2.

[3] Alberts, David S. and Richard E. Hayes. *Power to the Edge: Command and Control in the Information Age.* Arlington, Virginia: Command and Control Research Program, 2003

[4] Ambler, Scott. *The Object Primer.* 3rd ed. Cambridge: Cambridge University Press, 2004.

[5] Angehrn, Albert A., and Hans-Jakob Luthi. "Intelligent Decision Support Systems: A Visual Interactive Approach." *Interfaces*, 20:6 1990: 17-28.

[6] Appleton, Daniel, S. *Business Rules: the Missing Link.* Fairfax: D. Appleton Company, 1999.

[7] Arango, Guillermo, and Ruben Prieto-Diaz. "Introduction and Overview." *Domain Analysis and Software Systems Modeling.* Eds. Ruben Prieto-Diaz and Guillermo Arango. Los Alamitos, CA: IEEE Computer Society Press, 1991.

[8] Arbor Software, "On-Line Analytical Processing." Sunnyvale: Arbor Software 1997.

[9] Ashton, Robert H. *Human Information Processing in Accounting.* Sarasota: American Accounting Association, 1982.

[10] Ashton, Robert H., and Alison Hubbard Ashton, eds. *Judgment and decision-making research in accounting and auditing.* Cambridge: Cambridge University Press, 1995.

[11] Bass, Len , Paul Clements, and Rick Kazman. *Software architecture in practice.* Boston: Addison-Wesley Longman Publishing Co., Inc., 1998.

[12] Baum, David. "The Right Tools for Coding Business Rules." Datamation March 1, 1995: 36-38.

[13] Beale, Ian. Rev. of Why Information Systems Fail, by Chris Sauer. Internal Auditor August 1996: 12-14.

[14] Belkaoui, Ahmed. *Human Information Processing in Accounting.* New York: Quorum Books, 1989.

[15] Beltratti, Andrea, Sergio Margarita, and Pietro Terna. *Neural Networks for Economic and Financial Modeling.* London: International Thomson Computer Press, 1996.

[16] Ben-Ari, Mordechai. *Mathematical Logic for Computer Science.* 2nd ed. London: Springer, 1993.

[17] Bereny, Naveena. "Component Modeling with Rose 98." *Rose Architect* 1998.

[18] Bodnar, George H., and William S. Hopwood. *Accounting Information Systems.* Englewood Cliffs: Prentice Hall, Inc., 1995.

[19] Bohner, Shawn Anthony. "A Graph Traceability Approach for Software Change Impact Analysis." Diss. George Mason University, 1995.

[20] Booch, Grady, James Rumbaugh, and Ivar Jacobson. *The Unified Modeling Language Reference Manual*. Reading: Addison-Wesley, 1999.

[21] Booch, Grady; Rumbaugh, James; and Jacobson, Ivar. *The Unified Modeling Language Users Guide*. Reading, Massachusetts: Addison-Wesley, 1999.

[22] Booch, Grady; Rumbaugh, James; and Jacobson, Ivar. *The Unified Software Development Process*. Reading, Massachusetts: Addison-Wesley, 1999.

[23] Boritz, J. Efri. *CAPEX: A Knowledge-Based Expert System*. Princeton: Markus Wiener Publishers, 1996.

[24] Bowen, Jonathan P., and Michael G. Hinchey. "Ten Commandments of Formal Methods." *IEEE Computer* April, 1992: 56-62.

[25] Boxley, Maria, and Ahmad Wasim. "COOL:Plex Business Objects: Addressing New Generation of Techniques." *CBD Edge* November, 1999.

[26] Brachman, Ronald J., *et al*. "Mining Business Database

[27] s." *Communications of the ACM*, 39:11 1996: 42-48.

[28] Bradshaw, Jeffrey M. *Software Agents*. Cambridge: AAAI Press/MIT Press, 1997.

[29] Brandt, Dale O. et al. "R6Sigma Project Planning Workbook, NTB Reach Back." n.p.:n.p., May 2005.

[30] Brenne, Gordon. "Reengineering Management Controls to Improve Operations." *Association of Government Accountants Journal*, 44:4 1996: 32-37, 41-42.

[31] Brown, Alan W. "Understanding Enterprise Application Integration." *Component Online Strategies* November, 1999.

[32] Brown, Alan W., and Kurt C. Wallnau. "The Current State of CBSE." *IEEE Software* September/October, 1998.

[33] Brown, Carol E., and Mary Ellen Phillips. *Expert Systems for Management Accounting Tasks*. Montvale: The Institute of Management Accountants, 1995.

[34] Brown, Carol E., and Uma G. Gupta. "Applying Case-Based Reasoning to the Accounting Domain." *Intelligent Systems in Accounting, Finance and Management*, 3 1994: 205-221.

[35] Bruns, William J., Jr. *Accounting for Managers*. Cincinatti: South-Western Publishing Co., 1994.

[36] Buschmann, Frank, *et al*. *Pattern-Oriented Software Architecture A System of Patterns*. New York: Wiley & Sons, 1996.

[37] BusinessWeek online. "Online Banking Electronic Funds Transfer." April 2004: 1-2. April 2004 <http//partners.financenter.com/businessweek/learn/guides/onlinebanking/obresource.fcs>.

[38] Cantor, Murray Cantor. "Rational Unified Process for Systems Engineering: Part 1: Introducing RUP SE Version 2.0." *The Rational Edge* August 2003.

[39] Carey, Stephen S. *A Beginner's Guide to Scientific Method*. New York: Wadsworth Publishing Company, 1998.

[40] Carroll, John M., ed. *Scenario-Based Design*. New York: John Wiley & Sons, Inc., 1995.

[41] Case, Albert F., Jr. *Information Systems Development: Principles of Computer-Aided Software Engineering.* Englewood Cliffs: Prentice-Hall, 1986.

[42] Chan, Sally M. and Terence L. Lammers. "Reusing a Distributed Object Domain Framework." IEEE 1998.

[43] Chastek, Gary. "Object Technology and Product Lines." ACM Workshop 1997

[44] Chu, Pai-Cheng. "An Object-Oriented Approach to Modeling Financial Accounting Systems." *Accounting, Management & Information Technology,* 2, No. 1 1992: 39-56.

[45] Chu, Pai-Cheng. "Applying Object-Oriented Concepts to Developing Financial Systems." *Journal of Systems Management* May 1992: 28-34.

[46] Clarke, Roger. "Appropriate Research Methods for Electronic Commerce." March 2004:. April 2000. http://www.anu.edu.au/people/Roger.Clarke/EC/ResMeth.html,.

[47] Coad, Peter and Edward Yourdan. *Object-Oriented Analysis.* 2nd ed. Englewood Cliffs: Yourdan Press, 1991.

[48] Coad, Peter and Edward Yourdan. *Object-Oriented Design.* Englewood Cliffs: Yourdan Press, 1991.

[49] Cohen, Sholom and Linda M. Northrop. "Object-Oriented Technology and Domain Analysis." IEEE 1998.

[50] Cohen, Sholom G., *et al.* "Application of Feature-Oriented Domain Analysis to the Army Movement Control Domain." CMU/SEI June 1992.

[51] Colbert, Janet L. and Paul L. Bowen. "Comparison of Internal Controls: COBIT, SAC, COSO and SAS 55/78." *IS Audit and Control Journal,* IV 1996: 26-35.

[52] Colleen Frye, Colleen. "Application Integration & Management/Data Warehouse EAI Tools: Can They Cut IT consulting costs?" *Application Development Trends* April, 1999.

[53] Common Criteria Project Sponsoring Organisations. *Common Criteria for Information Technology Security Evaluation Version 2.1.* n.p: n.p, August, 1999.

[54] Computer Science and Telecommunications Board, National Research Council. *Realizing the Potential of C4I, Fundamental Challenges.* Washington, D.C.: National Academy Press, 1999.

[55] Curtis, Mary B. "Executive Information Systems." *IS Audit and Control Journal,* II 1995: 25-28.

[56] D'Souza, Desmond and Alan Wills. *Objects, Components and Frameworks with UML: the Catalysis Approach.* Addison-Wesley, 1998.

[57] Daniels, John. "Objects and Components: the Concepts." *CBD Edge* November, 1999.

[58] Davis, Alan M. *Software Requirements.* Revision. Upper Saddle River: Prentice Hall, 1993.

[59] de Borchgrave, Anrnaud. "Electronic bank robbers flourish." *The Washington Times* April 21, 1997.

[60] De Lucia, Andrea, et al. "Recovering Traceability Links in Software Artifact Management Systems using Information Retrieval Methods." *ACM Transactions on Software Engineering and Technology,* Vol 16, No. 4, Article 13, September 2007.

[61] Deitel, H.M. and P.J. Deitel. *C++ How to Program.* 3rd ed. Prentice Hall: New Jersey, 2000.

[62] DeMarco, Tom. *Structured Analysis and System Specification.* Englewood Cliffs: Yourdan Press, 1979.

[63] Department of Defense, Defense Information Systems Agency. *DII COE I&RTS: Rev 4.0 (Draft).* n.p.: n.p., 1999.

[64] Department of Defense. DoDAF Architecture Working Group. *DoD Architecture Framework Version 1.0.* August 2003.

[65] Department of Defense. *Joint Technical Architecture, Joint Interoperability and Warrior Support Version 5.0.* n.p.: n.p., April, 2003.

[66] Desouza, Kevin C. Rev. of *Managing Software Engineering Knowledge, eds.* A. Aurum, R. Jeffery, C. Wohlin, and M. Handzic. *Knowledge Management Research & Practice* 2 (2004): 63–64.

[67] Di Nitto, Elisabetta and Alfonso Fuggetta. "Product Lines: what are the issues?" *IEEE* 1997.

[68] Dill, David L., and Rushby, John. "Acceptance of Formal Methods: Lessons from Hardware Design." *IEEE Computer* April 1996: 23-24.

[69] Dilthey, William. *Survey of the System of the Particular Human Sciences, in which the Necessity of a Foundational Science Is Demonstrated.* Trans. Michael Neville. Braunschweig: Vieweg, 1896.

[70] Durlach, Nathaniel I. and Anne S. Mavor, eds. *Virtual Reality – Scientific and Technological Challenges.* Washington: National Academy Press, 1995.

[71] Edward, Alex and N.A.D. Connell. *Expert Systems in Accounting.* Englewood Cliffs: Prentice Hall Inc., 1989.

[72] Ehnebuske, Dave, *et al.* "Business Objects and Business Rules." *Business Object Workshop III, OOPSLA.* 1997.

[73] Eisenhardt, Kathleen M., (1989) *"Building Theories From Case Study Research," Academy Of Management Review,* 14(4) 532-550.

[74] Elliott, Robert K. "The Future of Assurance Services: Implications for Academia." *Accounting Horizons,* 4 1995: 118-127.

[75] Eriksson, Hans-Erik, *et al. UML 2 Toolkit.* Indianapolis: Wiley Publishing, Inc., 2004.

[76] Fayol, Henri. *General and Industrial Management.* 1916. Ed. Irwin Gray. Belmont, California: David S. Lake Publishers, 1987.

[77] Fernandez, Jose L. "An Architectural Style for Object Oriented Real-Time Systems." IEEE 1998.

[78] Foner, Lenny. "What's an agent? Crucial notions." 1993. April 2003. <http://foner.www.media.mit.edu/people/foner/Julia/Julia.html>.

[79] Fowler, Martin. *Analysis Patterns: Reusable Object Models.* Addison-Wesley: Reading, 1997.

[80] Francalanci, C. and A. Fuggetta. "Integrating Information Requirements." *Software Engineering Notes,* 22:1 1997: 57-68.

[81] Franklin, Stan and Art Graesser. "Is it an Agent, or just a Program?: A Taxonomy for Autonomous Agents." *Proceedings of the Third International Workshop on Agent Theories, Architectures, and Languages.* Springer-Verlag, 1996.

[82] Fraser, Steven, *et al.* "Application of Domain Analysis to Object-Oriented Systems." Addendum to the Proceedings OOPSLA '95 1995: 46- 49.

[83] Galloway, Duncan J. "Control Models in Perspective." *Internal Auditor* December 1994: 46-52.

[84] Gamma, Erich, *et al*. Design Patterns: Elements of Reusable Object-Oriented Software. Reading: Addison-Wesley, 1995.

[85] Garceau, Linda and Craig Foltin. "Neural Networks: A New Technology and the Impact on IS Auditors." *IS Audit and Control Journal*, II 1995: 52-57.

[86] Garlan, David, and Mary Shaw. "An Introduction to Software Architecture," *Advances in Software Engineering and Knowledge Engineering, Volume I*. Eds. V.Ambriola and G.Tortora. New Jersey: World Scientific Publishing Company, 1993.

[87] Gauntt, James E., Jr. and G. William Glezen. "Analytical Auditing Procedures." *Internal Auditor* February 1997: 56-60.

[88] Geers, Guido L. and William McCarthy. "Automated Integration of Enterprise Accounting Models Throughout the Systems Development Life Cycle." *Intelligent Systems in Accounting*, 5 1996: 113-128.

[89] Gelinas, Ulric J. and Alan E. Oram. *Accounting Information Systems*. 3rd ed. Cincinnati: South-Western College Publishing, 1996.

[90] Gessford, John. "Object-Oriented Cost Accounting System Design." *Journal of End User Computing*, 5:3 1993: 17-25.

[91] Glass, Robert L. "Formal Methods Are a Surrogate for a More Serious Software Concern." *IEEE Computer* April 1996: 19.

[92] Gleim, Irvin N. *CIA Review*. Vol. I. 6th ed. Gainesville: Gleim Publications, Inc., 1995.

[93] Gomaa, Hassan, *et al*. "A Knowledge-Based Software Engineering Environment for Reusable Software Requirements and Architectures." *INFT 803: Reusable Software Architectures: Course Readings*. Comp. and ed. Hassan Gomaa. Fairfax, Virginia: George Mason University, 1999.

[94] Gomaa, Hassan. "An Object-Oriented Domain Analysis and Modeling Method for Software Reuse." *INFT 803: Reusable Software Architectures: Course Readings*. Comp. and ed. Hassan Gomaa. Fairfax, Virginia: George Mason University, 1999

[95] Gomaa, Hassan. "Example of Domain Modeling Factory Automation Domain." *INFT 803: Course Notes on Object-Oriented Analysis and Modeling for Families of Systems*. Comp. and ed. Hassan Gomaa. Fairfax, Virginia: George Mason University, 1999.

[96] Gomaa, Hassan. "Reusable Software Requirements and Architectures for Families of Systems." *INFT 803: Reusable Software Architectures: Course Readings*. Comp. and ed. Hassan Gomaa. Fairfax, Virginia: George Mason University, 1999.

[97] Gomaa, Hassan. *INFT 803: Reusable Software Architectures: Course Notes on Object Oriented Analysis and Modeling*. Fairfax: George Mason University, 1999.

[98] Gomaa, Hassan. *INFT 803: Reusable Software Architectures: Course Notes on Object Oriented Analysis and Modeling for Families of Systems*. Fairfax: George Mason University, 1999.

[99] Gottesdiener, Ellen. "Turning Rules into Requirements." *Application Development Trends* July 1999.

[100] Gottesdiener, Ellen. "Business Rules Show Power, Promise." *Application Development Trends* March 1997: 36-52.

[101] Gower, Barry. *Scientific Method An Historical and Philosophical Introduction*. London: Routledge, 1997.

[102] Graham, Gal and William E. McCarthy. "Specification of Internal Accounting Controls in a Database Environment." *Computers & Security*. 4 1985: 23-32.

[103] Gregory, Donald R. *Mind Your Logic*. Dubuque, Iowa: Kendall/Hunt Publishing Company, 1996.

[104] Gries, David. "The Need for Education in Useful Formal Logic." *IEEE Computer* April 1996: 29-30.

[105] Gries, David. *The Science of Programming*. New York: Springer-Verlag, 1981.

[106] Griss, Martin L., *et al.* "Integrating Feature Modeling with the RSEB." IEEE 1998.

[107] Hall, Anthony. "Seven Myths of Formal Methods." *IEEE Software* September 1990.

[108] Hall, Anthony. "What is the Formal Methods Debate About?" *IEEE Computer* April 1996: 22-23.

[109] Hammer, Michael and James Champy. *Reengineering the Corporation*. New York: HarperBusiness, 1994.

[110] Harmon, Paul. "Sterling's COOL:Plex." *CBD Edge* November 1999.

[111] Harmon, Paul. "Turning Legacy Applications into Component-Based Systems." *Component Strategies* September 1999.

[112] Harvard University. "FCPA Fact Sheet." Office of the General Council. October 19, 2004.

[113] Hay, David, and Keri Anderson Healy. "Guide Business Rules Project Final Report." rev. 1.2. n.p.: GUIDE October 1997.

[114] Hempel, Carl G. *Aspects of Scientific Explanation*. New York: The Free Press, 1965.

[115] Heymann, H. G. and Robert Bloom. *Decision Support Systems in Finance and Accounting*. New York: Quorum Books, 1988.

[116] Hibbard, Justin. "Spreading Knowledge." *Computerworld* April 7, 1997: 63-64.

[117] Hinchey, Michael G and Jonathan P. Bowen. *Industrial-Strength Formal Methods in Practice*. London: Springer, 1999.

[118] Hinchey, Michael, and Bowen, Jonathan P. "To Formalize or Not to Formalize." *IEEE Computer* April, 1996, 18-19.

[119] Holloway, Michael C. "Impediments to Industrial Use of Formal Methods." *IEEE Computer* April, 1996, 25-26.

[120] Huhns, Michael and Munindar P. Singh, eds. Readings in Agents. San Francisco: Morgan Kaufmann Publishers, Inc., 1998.

[121] Hunter, Geoffrey. *Metalogic an Introduction to the Metatheory of Standard First d*

[122] IBM Rational Software. *Rational Rose RealTime Online Help*. Version 2003. <www.rational.com>.

[123] IBM Rational Software. "IBM Rational Rose RealTime: A guide for evaluation and review." June 2003:1-19. April 2004 < www-136.ibm.com/developerworks/rational/>.

[124] IBM. *IBM Dictionary of Computing*. George McDaniel, ed. 10th ed. McGraw-Hill, Inc., 1993.

[125] ILOG. ILOG Rules White Paper. Mountain View, California: n.p., 2001.

[126] ILOG. *Business Rules Powering Business and e-Business, White Paper.* Mountain View, California: n.p., 2001.

[127] Imielinski, Tomasz and Heikki Mannila. "A Database Perspective on Knowledge Discovery." *Communications of the ACM*, 39:11 (1996), 58-64.

[128] Immon, W. H. "The Data Warehouse and Data Mining." *Communications of the ACM*, 39:11 (1996), 49-50.

[129] Inman, E.E. "Enterprise modeling advantages of San Francisco for general ledger systems." IBM Systems Journal, vol. 37, no. 2.

[130] Institute of Internal Auditors, *Standards for the Professional Practice of Internal Auditing.* Altamonte Springs: Institute of Internal Auditors, 1978.

[131] Jackson, Daniel, and Jeannette Wing. "Lightweight Formal Methods." *IEEE Computer* April, 1996, 21-22.

[132] Jacobson, Ivar and Stefan Bylund. "A Multiagent System Assisting Software Developers." *Application Development Trends*, June 2002.

[133] Johnson, Ralph E. "Components, Frameworks, Patterns (extended abstract)." Software Engineering Notes, 22(3), May, 1997.

[134] Johnson, Ralph. "Business Objects." November, 1999. <st.t-www.cs.uiuc.edu/users/Johnson/bus-obj.html>, (November, 16, 1999).

[135] Joint Staff J3, Joint C4ISR Decision Support Center. *Joint Task Force (JTF) Command and Control (C2) Operational Concept Study, Phase 1 Final Report, 2001.*

[136] Jones, Capers. *Applied Software Measurement.* 2nd ed. New York: McGraw-Hill, 1996.

[137] Jones, Cliff B. "A Rigorous Approach to Formal Methods." *IEEE Computer* April, 1996, 20-21.

[138] Kandelin Nils A. and Daniel E. O'Leary. "Verification of Object-Oriented Systems: Domain-Dependent and Domain-Independent Approaches." *Journal of Systems Software*, 29 (1995), 261-269.

[139] Kandelin, Nils A. and Thomas W. Lin. "A Computational Model of an Events-Based Object-Oriented Accounting Information System for Inventory Management." *Journal of Information Systems*, Spring (1992), 47-62.

[140] Kang, Kyo C., *et al. Feature Oriented Domain Analysis (FODA) Feasibility Study.* Pittsburgh: Carnegie Mellon University, 1990.

[141] Kean, Liz. "Domain Engineering and Domain Analysis." *Software Technology Review.* New York: Software Engineering Institute, 1997.

[142] Ketz, J. Edward. *Bridge Accounting: Procedures, Systems and Controls.* Europe: Wiley, 2001.

[143] Kimbraugh, Steven O., *et al.* "The Coast Guard's KSS Project." *Interfaces*, 20:6 (1990), 5-16.

[144] Klein, Michel R. and Leif B. Methlie. *Knowledge-based Decision Support Systems.* New York: John Wiley & Sons, Inc., 1995.

[145] Kleppe, Annek, and Jos Warmer. "Extending OCL to Include Actions." UML 2000. Eds. A. Evans, S. Kent, and B. Selic. Berlin: Springer-Verlag, 2000, 440-450.

[146] Kleppe, Annek, and Jos Warmer. *The Object Constraint Language*. Reading, Massachusetts: Addison-Wesley, 1999.

[147] Koppelberg, Sabine. *Vol. 1 of Handbook of Boolean Algebras*. 3 vols. Ed. J. Donald Monk. New York: North-Holland, 1989.

[148] Kruchten, Philippe. "Planning an Iterative Project." *The Rational Edge e-zine for the rational community*. October, 2002.

[149] Kruchten, Philippe. *The Rational Unified Process*. Reading: Addison-Wesley, 1999.

[150] Kruchten, Phillippe. "Architectural Blueprints – the "4+1" Model of Software Architecture." *IEEE Software* November, 1995, 42-50.

[151] Kruse, Mark J. "Computers and Auditing." *Internal Auditor*, February, 1997,16-20.

[152] Kuhn, Thomas S. *The Structure of Scientific Revolutions*. 2d ed. Vol. 2, No. 2 of International Encyclopedia of Unified Science. Chicago: The University of Chicago, 1970.

[153] Larsen, Robert. "A continuous early validation method for improviing the Development Process." Diss. George Mason University, 2000.

[154] Laudan, Larry. *Science and Hypothesis*. London: D. Reidel Publishing Company, 1981.

[155] Levis, Alexander H. and Lee W. Wagenhals. *C4ISR Architecture Framework Implementation*. AFCEA Educational Foundation: Fairfax, 1999.

[156] Liskov, Barbara H. and Jeanette M. Wing. "A Behavioral Notion of Subtyping." *ACM Transactions on Programming Languages and Systems,* Vol 16, No. 6, November 1994, pp. 1811-1841.

[157] Lucente, Mark. "Computational holographic bandwidth compression." *IBM Systems Journal*, October, 1996.

[158] Lutz, Michael. "Consumable Mathematics for Software Engineers." *IEEE Computer* April, 1996, 27-28.

[159] Luz, Saturnino. "Software Agents: An overview." Trinity College, Department of Computer Science, February 2002.

[160] Lyons, Andrew. "UML for Real-Time Overview," Rational Software Inc. 1998.

[161] Mangina, Eleni. "Review of Software Products for Multi-Agent Systems." Applied Intelligence (UK) LTD, for AgentLink, June 2002.

[162] McCarthy, William E. "An Entity-Relationship View of Accounting Models." *The Accounting Review*, LIV, No. 4 (1979), 667-685.

[163] McCarthy, William E. "The REA Accounting Model: A Generalized Framework for Accounting Systems in a Shared Data Environment." *The Accounting Review*, LVII, No. 3 (1982), 554-578.

[164] McDermid, John A., ed. *Software Engineer's Reference Book*. Boca Raton: CRC Press Inc., 1993.

[165] Medvidovic, Nenad, *et al*. "Reuse of Off-the-Shelf Components in C2-Style Architectures." *ACM* (1997).

[166] Metamodel.com. "What is Metamodeling?" October 2003: 1-4. October 2003 <www.metamodel.com/staticpages/index.php?page=20021010231056977>.

[167] Meyer, Bertrand. *Object-Oriented Software Construction*. 2nd ed. Santa Barbara: ISE Inc., 1997.

[168] Morandin, Elisabetta, Gianfranco Stellucci, and Francesco Baruchelli. "A Reuse-Based Software Process Based on Domain Analysis and OO Framework." IEEE 1998, 890-891.

[169] Mowbray, Thomas J. "Architecture-Centered Development or Bust." *Enterprise Development* (Fall 1998).

[170] Mu¨ller-Merbach, Heiner. "Socrates' Warning Knowledge is more than information." *Knowledge Management Research & Practice* 2 (2004): 61–62.

[171] Murch, Richard and Tony Johnson. *Intelligent Software Agents*. Upper Saddle River, New Jersey: Prentice-Hall, Inc., 1999.

[172] Murthy, Uday S. and Casper E. Wiggins, Jr. "Object-Oriented Modeling Approaches for Designing Accounting Information Systems." *Journal of Information Systems*, 7:1 (1993), 97-111.

[173] Nagel, Ernest. *The Structure of Science Problems in the Logic of Scientific Explanation*. New York: Harcourt, Brace & World, Inc., 1961.

[174] National Aeronautics and Space Administration. "Technology Readiness and Roadmapping." np:np, April 2001.

[175] Neighbors, J. "Software Construction Using Components." Diss. University of California, Irvine, 1981.

[176] Nengsheng, Zhang, *et al*. "Embedding Knowledge-Based Systems into C++ Applications." *IEEE International Conference on Intelligent Systems for the 21st Century*, 4:3180-3185, October 1995.

[177] *Network Centric Warfare Report to Congress*. Department of Defense, July, 2001.

[178] Nissanke, Nimal. *Formal Specification Techniques and Applications*. London: Springer, 1999.

[179] Nwana, Hyacinth S. "Software Agents: An Overview." *Knowledge Engineering Review*, Vol. 11, No 3, September 1996, 1-40

[180] O'Leary, Daniel. *Expert Systems and Artificial Intelligence in Internal Auditing*. Princeton: Markus Wiener Publishers, 1995.

[181] OMG.org. April 2004. Object Management Group. April 2004. <www.OMG.org.>.

[182] Olle, William T., *et al*. *Information Systems Methodologies: A Framework for Understanding*. Wokingham: Addison-Wesley, 1991.

[183] O'Meara, Kelly Patricia. "Why Is $59 Billion Missing From HUD?" *Insight on the News*. December 27, 2004. December 27, 2004. <http://www.insightmag.com/news/208636.html>.

[184] Pande, Peter S., Robert P. Neuman, and Roland R. Cavanagh. *The Six Sigma Way*. New York: McGraw Hill, 2000.

[185] Pap, Arthur. *An Introduction to the Philosophy of Science*. New York: The Free Press, 1962.

[186] Parnas, D. "Designing Software for Ease of Extension and Contraction." IEEE Transactions on Software Engineering, March, 1979.

[187] Parnas, David Lorge. "Mathematical Methods: What We Need and Don't Need. *IEEE Computer* April, 1996, 28-29.

[188] Patriotta, Gerardo. "On studying organizational knowledge." *Knowledge Management Research & Practice* 2 (2004): 3–12.

[189] Petrie, <u>Charles J.</u> "Agent-Based Engineering, the Web, and Intelligence." *IEEE Expert*, December 1996.

[190] Plecki, Marian. *The Logic of Empirical Theories*. London: Routledge & Kegan Paul, 1969.

[191] Poincare, Henri. *Science and Hypothesis*. Trans. Francis Maitland. New York: Dover Publications, Inc., 1952.

[192] Poincare, Henri. *Science and Method*. Trans. Francis Maitland. New York: Dover Publications, Inc., 1908.

[193] Polanyi, Michael. *Personal Knowledge Toward a Post-Critical Philosophy*. Chicago: The University of Chicago Press, 1974.

[194] Polanyi, Michael. *The study of man*. Chicago: University of Chicago Press.

[195] Polanyi, Michael. *The tacit dimension*. Garden City, N.Y.: Doubleday, 1966.

[196] Pomeranz, Felix. *The Successful Audit – New Ways to Reduce Risk Exposure and Increase Efficiency*. Homewood: Business One Irwin, 1992.

[197] Potter, Ben, *et al. An Introduction to Formal Specification*. 2nd ed. London: Prentice Hall, 1996.

[198] Powers, Michael J., Paul H. Cheney, and Galen Crow. *Structured Systems Developmen.*, 2nd ed. Boston: Boyd & Fraser Publishing Company, 1990.

[199] Prawitt, Douglas F., and Marshall B. Romney. "Emerging Business Technologies." *Internal Auditor*, February, 1997, 25-32.

[200] Pressman, Roger S. *Software Engineering: A Practitioner's Approach*. 4th ed. McGraw-Hill Companies, Inc.: New York, 1997.

[201] Pressman, Roger S. *Software Engineering: A Practitioner's Approach*. 5th ed. McGraw-Hill Companies, Inc.: New York, 2001.

[202] Pressman, Roger S. *Software Engineering: A Practitioner's Approach*. 6th ed. McGraw-Hill Companies, Inc.: New York, 2005.

[203] Prieto-Diaz, Ruben. "Domain Analysis for Reusability." *Domain Analysis and Software Systems Modeling*. Eds. Ruben Prieto-Diaz and Guillermo Arango. Los Alamitos, CA: IEEE Computer Society Press, 1991.

[204] Quatrani, Terry. *Visual Modeling with Rational Rose and UML*. Reading: Addison Wesley, 1998.

[205] Ratliff, Richard L., *et al. Internal Auditing Principles and Techniques*. 2nd ed. Altamonte Springs: Institute of Internal Auditors, 1996.

[206] Rine, David and Raiek Alnakari, "A Four-Valued Logic B(4) of E(9) for Modeling Human Communication." Proceedings of the ISMVL2000 Conference, *IEEE Computer Society* April 2002.

[207] Rine, David and Kristina Statnikova. "Qualitative Research–Case Study." n.p.:n.p., April 2003.

[208] Robinson, Keith, and Graham Berrisford. *Object-oriented SSADM*. New York: Prentice Hall, 1994.

[209] Rule Machines Corporation. "Managing Business Rules: A Repository Approach." n.p.: n.p., 1999.

[210] Rumbaugh, James, *et al. Object-Oriented Modeling and Design*. Englewood Cliffs: Prentice Hall, 1991.

[211] Sage, Andrew P., *Decision Support Systems Engineering*. New York: John Wiley & Sons, Inc., 1991.

[212] Saiedian, Hossein. "An invitation to Formal Methods." *IEEE Computer* April, 1996, 16-17.

[213] Salmon, Merrilee H., *et al. Introduction to the Philosophy of Science*. Upper Saddle River, NJ: Prentice Hall, 1992.

[214] SAP-AG. Comprehensive Solutions: SAP Systems Offer Best Practices Solutions for Critical Business Processes. SAP-AG, 1999.

[215] Sahraoui, Abd-El-Kader and Issa Traore (2). "Experience with Multiformalism Strategies and Tools." n.p.: n.p., 2004.

[216] Saylor, Michael J., *et al.* "True Relational OLAP: The Future of Decision Support." *Database Journal*, November-December, 1995.

[217] Sayrs, Brian. "Architecting Multimedia Database Systems." *Object Magazine*, February, 1997, 52-57.

[218] Schlimmer, Jeffrey C. and Leonard A. Hermens. "Software Agents: Completing Patterns and Constructing User Interfaces." *Journal of Artificial Intelligence Research*, 1, 1993, pp. 61-89.

[219] SEI, Carnegie Mellon University. "Architectural Evaluation of Collaborative Agent-Based Systems." n.p.: n.p., 2003.

[220] Sessions, Roger. "Java 2 Enterprise Edition (J2EE) versus The .NET Platform Two Visions for eBusiness." May 15, 2001. <www.objectwatch.com> (June 10, 2003).

[221] Sharpe, Shane, *et al.* "Beyond Integrity Constraints." *Journal of Systems Management*, January-February (1996), 52-56.

[222] Shaw, Mary. "Writing Good Software Engineering Research Papers." *IEEE* 2003, 726-736.

[223] Siegel, Philip H., *et al. Applications of Fuzzy Sets and the Theory of Evidence to Accounting*. Greenwich: JAI Press Inc., 1995.

[224] Soileau, Jared S. "The Challenges and Effects of the Sarbanes-Oxley Act on the Internal Audit Profession." Louisiana State University, May 9, 2003.

[225] Spitzer, Tom. "Component Architectures." *DBMS and Internet Systems* September, 1997.

[226] Stedman, Craig. "Vendors try to cover all querying bases." *Computerworld* March 10, 1997, pp. 49, 52.

[227] Stedman, Craig. "Warehouse managers squeezed by user demand, limited systems." *Computerworld* February 3, 1997, pp. 45-46.

[228] Stedman, Craig. "Warehouse monitors in demand." *Computerworld* February 17, 1997, 43-45.

[229] Summers, Edward Lee. *Accounting Information Systems*. 2nd ed. Boston: Houghton Mifflin Company, 1991.

[230] Sunder, Shyam. "Failure of Controls: An E-Commerce Epidemic." *European Institute for Advanced Studies in Management: Workshop on E-Business and Management Controls*. Brussels, December 14, 2001.

[231] Sutton, Steve G., ed. *Advances in Accounting Information Systems*. Vol. 2. London: JAI Press Inc., 1992.

[232] Svoboda, Frank, *et al*. "Domain Analysis in the DoD." *ACM SIGSOFT Software Engineering Notes,* vol 21 no 1, January 1996, p. 57.

[233] Swanson, G. A. and Hugh L. Marsh. *Internal Auditing Theory: A systems View*. New York: Quorum Books, 1991.

[234] *Systems Engineering Fundamentals*. Fort Belvoir: Defense Systems Management College Press, December, 1999.

[235] Ta, Anh. "Technology Dependent Domain Modeling." Diss. George Mason University, 1998.

[236] Tellis, Winston. "Application of a Case Study Methodology." *The Qualitative Report*, Volume 3, Number 3, September, 1997

[237] Turban, Efraim and Jay E. Aronson. *Decision Support Systems and Intelligent Systems*, 5th ed. London: Prentice-Hall International, Inc., 1998.

[238] U.S. Army and the Defense Advanced Research Projects Agency, Government FCS Team, "FCS C4ISR IPT REPORT On Metrics, Functions, & Critical Issues." December 2000.

[239] U.S. Department of the Army. *Field Manual No. 7-10*. 14 December 1990.

[240] U.S. Department of the Army. *Field Manual No. 6-0* July 2003.

[241] Unknown Doctoral Dissertation. "An Empirical Analysis of REA Accounting Information Systems, Productivity, and Perceptions of Competitive Advantage." n.p.: n.p., [c. 1993].

[242] USoft. *Business Rules Automation*. n.p.: n.p., 1999.

[243] van den Brand, M.G.J., *et al*. "Reverse Engineering and System Renovation." *Software Engineering Notes*, 22:1 (1997), 57-68.

[244] Vasarhelyi, Miklos A. *Artificial Intelligence in Accounting and Auditing*. Vol. 3. Princeton: Markus Wiener Publishers, 1995.

[245] Vasarhelyi, Miklos A. *Artificial Intelligence in Accounting and Auditing*. Vol. 2. Princeton: Markus Wiener Publishers, 1995.

[246] Vasarhelyi, Miklos A. *Artificial Intelligence in Accounting and Auditing*. New York: Markus Wiener Publishing, Inc., 1989.

[247] Vici, Alessandro Dionisi, *et al*. "FODAcom: An Experience with Domain Analysis in the Italian Telecom Industry." IEEE, 1998.

[248] Von Halle, Barbara. *Business Rules Applied*. New York: Wiley Computer Publishing, 2002.

[249] Wang, Richard Y., *et al*. "A Framework for Analysis of Data Quality Research." *IEEE Transactions on Knowledge and Data Engineering*, vol. 7, no. 4, 1995.

[250] Warmer, Jos and Kleppe Anneke. *The Object Constraint Language: Precise Modeling with UML*. Addison-Wesley: Reading, Massachusetts, 1999.

[251] Weber, Ron. "Data Models Research in Accounting: An Evaluation of Wholesale Distribution Software." *The Accounting Review*, LXI, 3 (1986), 498-517.

[252] Weiss, Gerhard, ed. *Multiagent Systems*. Cambridge: MIT, 1999.

[253] White, Joseph B., *et al.* "Program of Pain." *The Wall Street Journal* [Eastern Edition], March 14, 1997, col. 6.

[254] Wickens, Christopher D. *Engineering Psychology and Human Performance.* New York: HarperCollins, 1992.

[255] Wiederhold, Gio. "Mediators in the Architecture of Future Information Systems." *IEEE Computer* March, 1992, 38-49.

[256] Wing, Jeannette M. "A Specifiers' Introductin to Formal Methods." *IEEE Computer* September, 1990, 8-24.

[257] Wohlin, Claes, *et al. Experimentation in Software Engineering: An Introduction.* Boston: Kluwer Academic Publishers, 2000.

[258] Wooldridge, Michael and Nicholas R. Jennings. "Intelligent Agents: Theory and Practice."

[259] Yannakopoulos, Demetrios, *et al.* "Object Lessons Learned from an Intelligent Agents Framework for Telephony-Based Applications." *Proceedings of the Technology of Object-Oriented Languages and Systems, Institute of Electrical and Electronics Engineers, Inc,* 1998.

[260] Yin, Robert (1994) *"Ch 1: Designing Case Studies," Case Study Research: Design & Methods*, 2nd edition, Thousand Oaks, CA: Sage Publications.

[261] Zachman, John A. "CONCEPTS OF THE FRAMEWORK FOR ENTERPRISE ARCHITECTURE Background, Description and Utility." La Cañada, California: Zachman International, 1993.

[262] Zahedi, Fatemeh. *Intelligent Systems for Business: Expert Systems with Neural Networks.* Belmont: Wadsworth Publishing Company, 1993.

[263] Zave, Pamela, and Michael Jackson. "Four Dark Corners of Requirements Analysis." *ACM Transactions on Software Engineering and Methodology*, 6:1 (1997), 1-29.

[264] Zave, Pamela. "Formal Methods Are Research, Not Development." *IEEE Computer* April, 1996, 26-27.

[265] Zelkowitz, Marvin V., and Dolores R. Wallace. "Experimental Models for Validating Technology." *IEEE Computing Practices* May 1998, 25-31.

[266] Zhao, Huimin. "Semantic Matching Across Heterogeneous Data Sources." *Communications of the ACM*, 50:1 2007: 45-50.

Wissenschaftlicher Buchverlag bietet

kostenfreie

Publikation

von

wissenschaftlichen Arbeiten

Diplomarbeiten, Magisterarbeiten, Master und Bachelor Theses
sowie Dissertationen, Habilitationen und wissenschaftliche Monographien

Sie verfügen über eine wissenschaftliche Abschlußarbeit zu aktuellen oder zeitlosen
Fragestellungen, die hohen inhaltlichen und formalen Ansprüchen genügt,
und haben **Interesse an einer honorarvergüteten Publikation**?

Dann senden Sie bitte erste Informationen über Ihre Arbeit per Email
an info@vdm-verlag.de. Unser Außenlektorat meldet sich umgehend bei Ihnen.

VDM Verlag Dr. Müller Aktiengesellschaft & Co. KG
Dudweiler Landstraße 125a
D - 66123 Saarbrücken

www.vdm-verlag.de

www.ingramcontent.com/pod-product-compliance
Lightning Source LLC
LaVergne TN
LVHW022300060326
832902LV00020B/3181